The Sky Is Home

The Sky Is Home

The Story of Embry-Riddle Aeronautical University

by

John McCollister
&
Diann Davis

 Jonathan David Publishers, Inc.
Middle Village, New York 11379

THE SKY IS HOME

Jonathan David Publishers, Inc.
68-22 Eliot Avenue
Middle Village, New York 11379

2 4 6 8 10 9 7 5 3 1

Library of Congress Cataloging-in-Publication Data
McCollister, John and Davis, Diann.
    The sky is home / by John McCollister & Diann Davis. — [2nd ed.]
      p.   cm.
    ISBN 0-8246-0386-9 (hardcover)
    1. Embry-Riddle Aeronautical University—History.
    I. Title.
TL560.2.E49M38 1996
629.13'0071'175919—dc20   96-5562
                                           CIP

Book design by Marcy Stamper

Printed in the United States of America

Acknowledgments

The family of Embry-Riddle Aeronautical University applauds the conscientious efforts of all those who compiled, wrote, and edited the written and oral testimonies about the University's foundation and development.

Ms. Diann Ramsden, former librarian for the University, gathered material from the Embry-Riddle archives and through visits to the Cincinnati Historical Society, the newspaper department of the Hamilton County (Ohio) Library, and Lunken Airport.

Dr. John McCollister wrote the narrative based upon this research as well as additional documents and oral testimony.

Finally, those who worked on this project thank their families for their support during the creation of this long-awaited story of Embry-Riddle.

*"For most people,
the sky is the limit;
for those who love aviation,
the sky is home."*

—JACK R. HUNT

Table of Contents

Introduction

This is the story of a university.

The word "story" is purposely used in lieu of the common designation "history." An institution of higher education rarely has a history in the strict sense of the word. Instead, as bestselling author Robert Serling (an honorary alumnus of Embry-Riddle) suggests, the record of its foundation and growth more closely resembles a biography. A private school, perhaps more than any other entity, resembles a human personality— usually because each campus mirrors the character of the individuals who created, nurtured, and guided it.

While many institutions never got beyond the planning stage, those that were able to begin and survive did so amid fierce competition from state schools, internal power struggles, and daring decisions. In spite of all the barriers thrown up to check their advance, the successful private colleges and universities share two legacies: bold leaders and priceless loyalty to the school.

In 1701, for instance, 10 Connecticut clergymen met in the village of Branford and donated books to found a college. In addition, the clergymen outlined the guiding principles that still serve as the foundation for what is known today as Yale University.

Likewise, in 1875, the dynamic and forceful Mormon leader Brigham Young shaped a university that bears his name so that students could study the dogma of their faith without fear of persecution.

Embry-Riddle Aeronautical University shares much of this same spirit, but with a few unique twists. Were Embry-Riddle to embody the distinct personalities of all the leading characters in

its development, it would create a psychologist's nightmare. The leaders who shaped its history form a collage of contrasts. Begun by two men—one whose main love was flying, the other whose primary philosophy was to enjoy life to the hilt—Embry-Riddle's standardbearers included the nation's first woman president of a flying school, a man who was forced to resign after serving less than six months, a former U.S. Navy commander who won the Harmon International Trophy, yet had no prior experience as a school administrator, a three-star general who once headed the U.S. Air Force Academy, and a Ph.D. who took a 70-percent cut in pay earned in the business world to lead the University into the 21st century.

The saga of Embry-Riddle is peppered with sharp peaks and valleys. The recorded memories portray a roller-coaster ride of prosperity and demise. The once-prosperous flying school literally closed its doors for nine years, only to be reborn when America was on the brink of war. After a decade marked with relocations and name changes, Embry-Riddle's administration dared to make some bold decisions. As a result, enrollments steadily increased, course offerings were expanded, and a university was born. Those who guided Embry-Riddle possessed the robust spirit of individualism so common within the aviation community. Yet, somehow, those egos were suppressed—often at the urging of a more powerful ego—and blended into a unified effort to mold the University into the world's leader in aviation/aerospace higher education.

This, then, is the story of Embry-Riddle and of the men and women dedicated to the proposition that students must be prepared to face every challenge aviation has to offer—from landing a single-engine biplane on a tiny grass strip in Cincinnati to launching a space shuttle from the pad at the Kennedy Space Center.

◄ **An Embry-Riddle Waco biplane in 1927. Pictured standing behind one Waco biplane of the "E-R Express" are** *(from left to right)* **John Paul Riddle, Durward "Duke" Ledbetter, T.H. Embry, Charlie Myers, John Woods, and Harry Sherwin.**

1

I Wanted to Fly

*"The Air Corps tried to send me
to balloon school. But I didn't
want to learn anything about balloons.
I wanted to fly airplanes."*

— J. PAUL RIDDLE, 1920

John Paul Riddle

He was only 18 years old, but he knew what he wanted out of life. One thing he knew he *didn't* want was a full-time military career. In fact, he initially declined the coveted appointments to both the U.S. Naval Academy and to West Point, although he finally accepted an appointment to Annapolis.

His name was John Paul Riddle. The year was 1920—only 17 years after the Wright Brothers flew their "Flyer" on its initial voyage of 12 seconds, covering all of 120 feet.

It was the beginning of a decade described by Isabel Laighton as "The Aspirin Age." The Twenties was an era when the Charleston and Eliot Ness were in vogue. People sang "Yankee Doodle Dandy." The stock market was going up; women's skirts were going down. Flappers, gangsters, and bathtub gin were paraded across the nation to the beat of a brash new sound of music called "jazz." Americans in the "Roaring Twenties" celebrated life with gusto. After all, they had just fought and won the "war to end all wars."

Aviation was still in the kindergarten of development. Aerial antics were introduced to the general public at county fairs by former Army pilots who barnstormed their way from town to town in war-surplus planes purchased for 800 dollars.

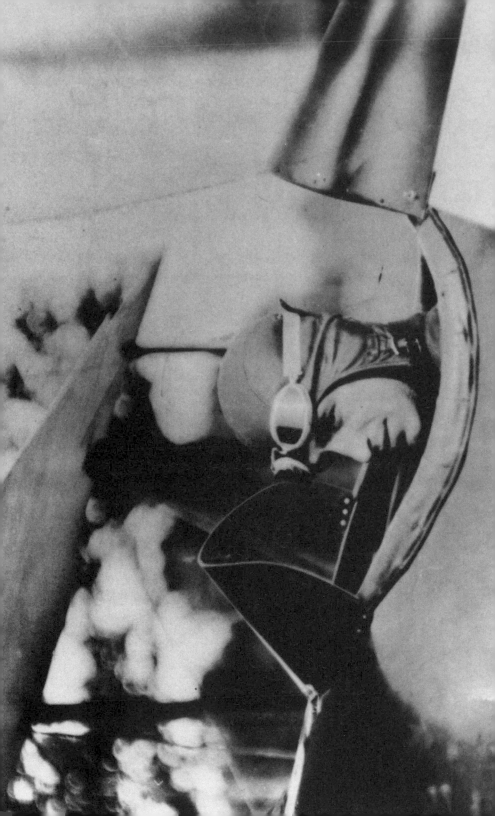

Air shows of these "flying gypsies" were staged to lure the more daring spectators on "joyrides" at 10 dollars for one trip "around the patch."

Although unorganized and expensive, aviation's mystique lured those who dared to rise above the earth.

John Paul Riddle was one of those people.

"I tried to join the Air Corps as a cadet as a way of learning to fly," said the six-foot, 140-pound native of Pikeville, Kentucky, "but, because of budget restrictions, they could not send me to flying school. They could send me to balloon school. However, I didn't want to learn anything about balloons. I wanted to fly airplanes."

Finally, the Air Corps agreed to send Riddle to its Air Service Mechanics School at Kelly Three Field in San Antonio, Texas.

"I spent a solid year learning how to build, repair, and maintain airplanes, engines, and accessories," he confessed, "and it was the best thing that ever happened to me. When I started taking planes into the air, I knew what made them tick. If something went wrong, I knew what it was, and I could fix it."

He completed mechanics school in 1921 and received his appointment as a cadet. He moved with the mechanics school from San Antonio to Chanute Field in Rantoul, Illinois, then to Carlstrom Field in Arcadia, Florida, for flight training.

He and his fellow cadets were limited to only five flying hours a month following their primary training, due to budget restrictions. Other cadets, however, seemed to be interested not so much in flying, but in what Sarasota, Ft. Myers, and other nearby Florida cities had to offer. They willingly relinquished their unused flying hours to Riddle who, as a result, accumulated more flying time than any other cadet on the field.

◀ **John Paul Riddle in the cockpit of an early biplane. Riddle was a cofounder of Embry-Riddle Aeronautical University.**

Soon he transferred to Post Field at Fort Sill, Lawton, Oklahoma, where he would learn, firsthand, about the value of sound construction. The Air Service there was rebuilding de Havilland DH-4s with Liberty engines. They were attached to an airframe unceremoniously dubbed "The Flying Coffin" because it could not withstand the pressures of aerobatics.

"On my first day there, my instructor took me up in one of those de Havillands," Riddle explained. "He takes me up about five thousand feet and says to me: 'Spin me down to thirty-five hundred.'"

"What?" I asked.

"Go on. Spin it," he said.

"Really spin it?"

"Yes," he demanded.

"So, I did. Nothing happened. I pulled it out at thirty-five hundred feet. It was smooth as silk."

"See? They make 'em a lot stronger and safer now."

"That," said Riddle, "showed me the importance of solid engineering."

Riddle received advanced training as an observation pilot when his entire observation unit was moved to Kelly Field, Texas. Both Riddle and one of his classmates, George McGinley—a former All American from Nebraska—wanted to share the adventure with others. Together they were able to muster $250 and purchase a used Jenny aircraft.

On weekends, they took off from Kelly Field to parts unknown, barnstorming around the state, giving rides to passengers for five or six dollars. "Sometimes pretty girls flew for free," said Riddle.

"I kept the plane hangared at Kelly Field," remembered Riddle. "My commanding officer was afraid that I might get hurt, so he ordered his operations officer to see that the plane was in number one condition at all times. Heck, they did all the work taking care of it. I had all the fun flying it."

Not only were his barnstorming trips profitable, but after his stint with the military, Riddle was also earning a solid reputation as a pilot and an instructor.

Lewis Stone, a former neighbor who had more money than experience, owned two airplanes and was instructing about 10 students. His fees were quite high—$300 for 10 hours of flying. At the same time, he was not a successful teacher.

"I'm having problems," he confessed to Riddle. "I get up with these students, and the plane starts to go over, and I can't properly control it."

"How much time have you had?" asked Riddle.

"Well...I went down to Huntington, West Virginia, and bought this airplane. They gave me one solo flight, then sent me home. I guess I have...ah...eight and a half hours. Three of them have been instructing."

"I couldn't believe it," Riddle recalls. "No wonder so many kids were killed in those days."

Over the next few months, Riddle instructed both Stone and his students, stressing the importance of safety—a dimension of aviation that he maintained could not be emphasized too much. Gangsters were labeled "public enemies" by law enforcement officials; to John Riddle, Public Enemy Number One was "pilot error," a killer which he fought like a G-man policing the still-dangerous sport of aviation.

Although safety was a prevailing concern, Riddle was not above employing a flair for the dramatic. Once when a passenger hired him to fly to Cincinnati, Riddle chose to land on a polo field that was close to the passenger's destination. Unfortunately, a game was in progress that Sunday afternoon. Riddle continued to circle the field, until both sides called a time-out between chukkers. Riddle promptly set the plane down on the field in front of some mildly disturbed spectators who thought the game might have to be postponed. Riddle maneuvered the plane to the sidelines and, while waiting for his passenger to return, demonstrated his uncanny ability to turn a problem into a profit.

Instead of arguing with the people, he convinced some of them to experience the thrill of flight...for a fee, of course.

Those who accepted his invitation became instant evangelists for flight. Many lost themselves in bold gestures and superlatives while describing to their friends and neighbors the magnificent view of life from a new perspective. Consequently, others wanted to share the thrill. Riddle enrolled enough students that afternoon to justify operating a small passenger-carrying service in Cincinnati. When business lagged, he returned to Pikeville and barnstorming, giving exhibition flights for county fairs and other celebrations.

Riddle predicted that Cincinnati was going to be a forerunner in aviation. In August 1923 he bought a Hisso Standard plane and set up a flying business at Besuden Field near Carthage. The business boomed until cold weather intervened. They followed the sun to Ashland, Kentucky, for more barnstorming, then continued on to more than a dozen states.

"When business was dull, we cut the price," admits Riddle. "When it was good, we raised it as high as we could go. In the eyes of these people, the 'Red Baron' had nothing on us. We were genuine heroes. We pocketed the money and basked in glory. That was my introduction to aviation management."

But financial income generated from barnstorming was, at best, sporadic. Riddle was eventually lured away by what appeared to be a more secure position with Griswold Air Service in Erie, Pennsylvania. After only a few months, however, he

▶ T. Higbee Embry, co-founder of the Embry-Riddle Company, left the enterprise in 1930 and moved to California, where he died in 1946. Embry and Riddle remained close friends through their pioneering partnership and afterward.

returned to Cincinnati to meet with a former student who would have a profound effect on his life and career.

The former student was T. Higbee Embry, a rather wealthy free spirit who suggested that the two of them become the distributors for Waco Aircraft in Ohio, Indiana, and Kentucky. With the signing of an agreement in December 1925, the Embry-Riddle Company was formed.

Talton Higbee Embry

T. Higbee Embry, a native of Cincinnati, first met Riddle in 1923. Riddle was then operating a small flying business at Besuden Field, which included selling rides to anyone who came by. One of the first to approach Riddle was Mr. Embry.

"How much for a ride?" asked Embry.

"How much do you have with you?" responded Riddle.

"Twenty dollars," he said.

"That's exactly the price. Hop in," said Riddle.

The rather expensive airfare (in that era) notwithstanding, Embry did not complain. Indeed, he was hooked. Within a few days, he bought his own airplane—a Waco 9 (the first one produced by Advance Aircraft Company)—and hired Riddle for flying lessons.

Embry was an excellent student. He soloed in only seven hours. Both Riddle and Embry developed a strong relationship during these days of instruction that grew beyond the teacher-student roles to a mutual respect for each other. Embry became convinced that his enthusiasm for life, combined with Riddle's flying expertise, would be the ideal raw materials for a profitable business in aviation. Riddle agreed.

On December 17, 1925—the 22nd anniversary of the Wright Brothers' first flight—Embry handed Riddle the agreement and a pen with which to sign. Everything was now in place for the actual start of business in 1926.

The Embry-Riddle Company

The Embry-Riddle Company began with two Jenny airplanes and a dream.

Its first venture was selling Waco Aircraft. The contract with the manufacturer guaranteed that a total of 12 airplanes would be delivered at one-month intervals. Embry and Riddle agreed to sell them as quickly as possible and to notify the manufacturer when the order needed to be increased.

Waco fulfilled its promise. Five planes were delivered over in the next five months, right on schedule. Unfortunately, Embry-Riddle could find no buyers.

"I felt so sorry for myself and Riddle," said Embry. "My heart sank when I saw those five Wacos sitting silently in the hangar just staring at us." Embry dug into his savings and bought one himself; his mother, now an officer of the Company, bought another.

The next few weeks were more rewarding. Both Embry and Riddle became more aggressive, combed the countryside, and located four buyers. Before the year ended, they sold a grand total of 14 aircraft.

"We knew we could do it, although there were moments at the beginning when we had it plenty rough," Embry confessed. "Others may have thrown in the towel, but that thought never crossed our minds. We were determined to make it. People often asked me if there was money in aviation. I told them, 'You're damn right. I ought to know. I put it there.'"

With the sales came both profits and other contracts. The Company expanded its operation at Lunken Airport in Cincinnati.

The Cincinnati Post applauded the growth:

> Regular commercial aviation service for Cincinnati is available as a result of an agreement between Army officials and controllers of the Embry-Riddle Company, airplane operators, which provides for joint occupancy of Lunken Airport.

The Lunken Airport was a genuine "home" to Riddle. "This was the same polo field in which I landed on my first flight to Cincinnati three years earlier," he said.

Other Ventures

Even before expanding at Lunken airport, the Company had inaugurated a new service—a flying school. A prominent businessman named John Pattison was the first student to solo. He was followed by scores of men and women who dared to leave the bonds of earth for a "trip through the heavens," as some of the more poetic enthusiasts phrased it.

Other services soon followed. "One of our first 'odd jobs' was to drop circulars for a Covington, Kentucky, store. They paid us two-hundred fifty dollars just for doing that," said Embry. "It was extremely windy that day. I remember when we dumped the circulars, they flew for miles. I wonder how many of them actually landed in Covington."

When Embry and Riddle chose to participate in the famous "Ford Reliability Tours"—a promotion sponsored by Edsel Ford to demonstrate the practicability of cities having airports for this fast-growing method of transportation—they made a unique impression. Riddle not only flew the Waco owned by Mrs. Susan Embry (Higbee's mother), but also took her along as a passenger.

She was the only woman on tour that year.

This was no publicity stunt. Riddle openly encouraged women to share in the thrill of flight.

◄ **Embry-Riddle Company Office in Cincinnati, Ohio. The company headquarters were located at the Lunken Airport from 1926 to 1930, when the airport and the company were considered "synonymous."**

"Women have more nerve than men when it comes to flying," claimed Riddle. "They are always more eager to go aloft, seem to enjoy it more than men, and invariably want to go up again and again. Maybe it's because women are naturally nearer to angels than are men."

The Company received national recognition when one of the pilots loaded film taken of the 1928 Kentucky Derby in Louisville and flew it to Ohio in time to load it aboard the train for New York for use by the International News Service the next day.

Neither Embry nor Riddle shunned publicity. They knew that a well-placed story could generate more business. On one occasion they asked the students of the University of Cincinnati to select a classmate whom Riddle's instructors would teach to solo in only one day.

"We did it just to show that learning to fly was not as difficult as many came to believe," said Riddle.

The student, Frank Sheldon, arrived at the airport at precisely 9:30 the next Saturday morning. Five and a half hours later, after an incredible number of 74 landings, young Sheldon took the plane around the field and brought it back with a perfect "three-point landing."

Embry and Riddle used every opportunity to trumpet the joys of flying. They, along with other employees of the Company, formed a "speakers' bureau" and appeared on a weekly radio show on Station WLW to answer questions about aviation.

While the company's pioneers never overlooked an opportunity for a good story, they never compromised on the importance of safety. Newspaper reporters of the day regularly used words such as "cautious" and "extremely careful" when describing the Embry-Riddle operation. One reporter wrote about Paul Riddle: "If he lives to be 120 years old, he will undoubtedly retain all of his fingers and toes because of his steadfast refusal to take unnecessary hazards."

▶ Flight instructors and air mail pilots. Photos taken in March 1929.

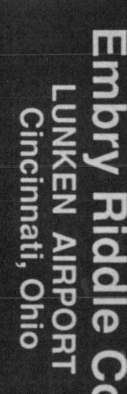

Air Mail Pilots

Embry Riddle Co.

LUNKEN AIRPORT

Cincinnati, Ohio

W. R. VINE

C. C. WEHRUNG

F. MERRILL

T. J. HILL

The constant awareness of safety was a theme trumpeted by Embry, Riddle, and every member of the Company team. Their persistence did not go unrewarded. The Company was commended in 1927 for flying 330,000 miles without a personal injury.

In December of that year, Contract Airmail Route 24, from Cincinnati, via Indianapolis, to Chicago, was awarded to Embry-Riddle. Several factors contributed to this selection as reported in Cincinnati's *Commercial Tribune:*

> The personnel and character of all firms bidding on air mail are vital considerations in the awarding of contracts. These facts were factors in the selection of the Cincinnati firm. Although the youngest, being but two and a half years old, it has already gained a name for itself throughout the industry by its aggressiveness, the quality of its flying school, and the high tone of operation maintained at Lunken Airport.

The mail service began that year on December 17—the date made famous in 1903 by the Wright Brothers' first airplane flight—and was ushered in with the red carpet treatment by the city. *The Commercial Tribune* carried a two-page spread; a motor parade escorted the mail truck to the airport; federal authorities, city officials, merchants, and professional men made certain they had front row seats. At 3:30 p.m., pilot Stanley "Jiggs" Huffman took off in the Waco painted a bright orange and silver (the adopted colors of the Embry-Riddle fleet) with a cargo of 7,200 letters and packages, while the onlookers cheered and applauded.

Cincinnati thus became one of the few cities to have direct mail service. The West Coast and the South were now three days closer.

Not only did Embry-Riddle win in competitive bidding, its performance record was so high that it was one of only two companies (the other was Boeing Air Transport) authorized by the Government to fly passengers as well.

Robert Serling, in his book *Eagle, the Story of American Air-*

lines, reports that passengers who flew with the mail were also pioneers of a sort. Prior to take-off, they were given a small box of provisions that might be deemed the first in-flight service. Each box contained an apple, a small sandwich, chewing gum, some cotton for the ears, and two small ammonia capsules—the standard treatment for airsickness.

Passengers did not enjoy the luxury of choice between first-class and coach accommodations. Instead, each one was required to sit on mailbags in the open cockpit in front of the pilot.

For some of the passengers, this was their introduction to flight. It was not for the fainthearted. Flying was still regarded by most as a daredevil stunt. Those brave enough to climb aboard what many considered to be little more than "fabric-covered deathtraps" and fly cross-country often boasted of their prowess for months thereafter.

Charles Planck, Embry-Riddle's first public relations officer, suggested a psychological aid for the more timid. He asked the pilots to carry two boxes of dirt on the floor of the front cockpit into which passengers could rub their feet and feel the crunch of earth. It was a gimmick, perhaps, but it had astounding results.

Pilots who flew these routes for Embry-Riddle would rather fly than eat. They were paid at the rate of only 10 cents a flying mile, with no base pay, and no differential for night flying. The pilots did not object, however, since the routes were over level ground, an important safety consideration.

Embry-Riddle, like every other mail carrier, was compelled to count every nickel. Air mail did not provide them with an avalanche of profits. The Government paid a maximum of three dollars for every pound of mail flown. When there wasn't enough mail to cover the cost of gasoline for the trip, some of the more inventive carrier pilots stuffed mail sacks with wet blotters, wrapped bricks, even bags of dirt—all with mailing labels, of course.

Air mail pilots flying cross-country often relied upon the crudest instruments—a compass, an altimeter, a watch, and the seat of their pants.

Their most often-used landmark was a highway, a telephone line, or a railroad track (dubbed an "iron compass"). When pilots were "on the rails," and a tunnel suddenly popped up out of nowhere, it could spell disaster. With no radio beam to guide them and no accurate way to measure ground speed, pilots who flew above the overcast could use only dead reckoning to know where they were or when to descend.

A rather creative Army pilot, Jimmy Doolittle, was working on a more reliable solution to the problem. He was engaged at McCook Field in the center of Dayton in the experiments that resulted in his famous totally "blind flight" one year later at Roosevelt Field, New York. Doolittle and his staff passed on these findings to J. Paul Riddle who, in turn, shared them with his students. The School used this new information to modify one of its training planes. Thus, Embry-Riddle became one of the first institutions to offer instrument flight instruction.

In just two years, Embry-Riddle had grown from a little-known airplane distributor to a multi-dimensional aviation enterprise receiving national attention.

One cannot help but wonder if T. Higbee Embry even dared to dream about this much success on that memorable afternoon in 1926 when his heart sank at the sight of those five unsold aircraft gathering dust in the hangar.

2

From School
to Airline

*"We start into a new year which we know
will be crammed full of history
in aviation—packed with events and
progress that we can't even imagine."*

— T. HIGBEE EMBRY, 1929

A Growing Reputation

Mr. Embry's fears had long since given way to smiles. By 1928, the Embry-Riddle Company owned six planes for the flying school, eleven for the air mail service, and two for cross-country flying. Fifty-four people, including 30 pilots, were on its payroll. In addition, the Company was now a sales agent not only for Waco but also Monocoupe and Fairchild aircraft.

Since its organization, it had sold 149 airplanes, which presented a new, rather pleasant problem—the Company had to find ways to get faster deliveries from the factories to keep pace with the orders.

The air mail business increased a whopping 112 percent. The Embry-Riddle air mail line was the only one in the country to show a 100-percent increase in poundage, according to one report. One of the reasons for this phenomenal growth was that the Company maintained its schedule 94 percent of the time.

"That was a hell of a lot better than even the slower trains," said Embry.

Through all of this increase and solid record of punctuality, the Company never forgot the importance of air safety. From 1927 to 1929, more than 22,000 passengers flew under the colors

of the Embry-Riddle Company with not one being injured. Punctuality and safety were noble qualities, indeed, but if aviation were to become attractive to the public, innovative appeals had to be used. Paul Riddle was a master at creating new ideas that caught the public's imagination.

In December 1929, Riddle offered flying lessons as Christmas gifts. These certificates were sold not at airports, but from the counters of department stores, cigar stores, and gift shops.

"If aviation expected to thrive," he said, "we had to reach out to where the people were."

The plan worked. Parents bought rides for their children. Pilots bought certificates for reluctant neighbors who swore: "Someday, I'm gonna' go up in one of those planes."

Air shows and air races of this era brought crowds of more than 50,000 people to airports on Sunday afternoons. Often the crowd at Lunken Field in Cincinnati surpassed those at any sporting event in the nation that day—including those who watched Babe Ruth and the mighty Yankees in New York.

Another means of reaching out to the general public was through the Air Information Bureau begun by the Company in 1928. Some historians claim that this was the world's first travel agency. The Bureau, located at the Gibson Hotel in Cincinnati, offered information on every air mail and passenger line in the world that operated on a given day. Through the Bureau, customers could find the best routes, the rates, and connecting bus or train schedules.

"We took care of all the details," said Riddle. "All the people had to do was bring their suitcases and follow our directions. It was as simple as that."

Simple, perhaps, but no one had ever thought of it before.

◄ **Two Monocoupes and a biplane. Three of the Embry-Riddle fleet in the late 1920s.**

The idea caught on. Later that year, Embry-Riddle, in cooperation with other air mail carriers, opened a similar operation, called "The Consolidated Airline Tickets Office," at the Palmer House in Chicago.

Safety

Due, in part, to the success of the Embry-Riddle Company, other flight schools were opened throughout America. Unfortunately, some seemed more concerned about turning a fast profit instead of offering quality instruction with safety.

One outspoken advocate for safety was Lt. Barrett Studley of the U.S. Navy who wrote: "The operation of these schools is of vital concern, not only to the student, but to everyone interested in aviation." The lieutenant explained: "For every student who, half-trained and ill-prepared, wrecks his plane on an early solo flight, ten other prospective airplane buyers are discouraged." The airplane manufacturers listened to Lt. Studley's warning with more than just passing interest. The number of future sales would be in direct proportion to well-trained new pilots.

Lt. Studley, in other articles and speeches delivered throughout the country, took exception to the poor quality of instruction offered at many schools. "A flight instructor," he said, "must be able to analyze mistakes, determine the causes, and explain them to the student so that he will understand. The responsibility of the flight schools is to share effective methods to not only teach flying, but also to correct or avoid common mistakes."

"Not all men can fly safely," he continued. "A flight school cannot escape the responsibility of separating those who can from those who cannot. It takes ten hours of dual flight to determine if someone has a satisfactory aptitude for this unforgiving concept called 'flying.'"

In contrast to the schools that were the targets of the lieutenant's wrath, Embry-Riddle enjoyed a solid reputation for teaching flying skills with safety.

Mr. Embry explained: "There are two types of commercial aviation—the 'aviation enterprise' and the 'aviation racket.' The 'aviation racket' pictures a well-dressed pilot with limited knowledge, flying a brightly-painted airplane with loose fabric and a noisy engine. This 'racket' is one real obstacle to the progress of aviation. Embry-Riddle will have no part of an 'aviation racket.'"

The editor of *Air Transportation* agreed when he wrote in 1928: "Every once in a while, someone comes along and develops an air school to the point where the school means a lot more than the school of earlier days. Embry-Riddle is to be congratulated for the high standards on which they conduct their school."

Another writer reported: "The Embry-Riddle Flying School is not only one of the best in the country, but is one of the best equipped."

The high standards were embodied in the philosophy of the School as spelled-out in its first official catalog in 1928:

- We believe that the student should learn in a school employing only licensed pilots and modern licensed planes.

- It should have financial and moral standing and stability in back of every statement or promise. It should have and merit the good will and respect of the entire industry.

- It should provide a thorough ground school training, with competent instructors, so that there will be an adequate background of the theory of flight to form the basis of a student's advancement in the profession of flying or to enhance his skill at the controls in flying for sport or pleasure.

- It should adhere strictly to the rules of the U.S. Dept. of Commerce, Aeronautics Branch, governing flying schools, both in spirit as well as technically.

- It should be conscientious, its diplomas presented for evidence of skill and knowledge attained only, and not for the mere cost of tuition.

The School's first formal commencement exercise conducted on the flight line, July 29, 1928, featured Stanley "Jiggs" Huffman, Operations Officer of the School, who delivered the commencement address. He, too, took this opportunity to stress the importance of flight safety: "Learn your limitations," he warned the students. "Know when to quit. People may say about you, 'That so-'n-so is a mediocre pilot.' They might say, 'You are yellow.' Don't worry. You would rather hear them say these things than to hear them say, 'He WAS a good pilot.'"

Flying as a Profession

Embry-Riddle had the unique position of contributing to the development of aviation as an employer and as a trainer of pilots. The Company was constantly on the lookout for pilots for its rapidly growing air passenger, air mail, and air express services, not to mention airplane sales.

As an employer, it recognized the responsibility in training students to meet its own demands, plus those of the industry.

Charles Planck wrote in a 1929 issue of *Aviation* magazine:

> The [Embry-Riddle] Company has realized the imperative needs of pilots, and is constantly on the lookout among its students for likely material for its own pilot force.

"We were at the point where we could be more selective than others," admitted Riddle. "We chose to make professionals out of those who had the right material and attitude."

One instructor put it more bluntly: "If they are the right type, we urge them on; if not, we drop them, because there are a lot more waiting."

More were waiting. From October 1928 through January 1929, the School averaged 100 letters of application each week.

Riddle's standards were known to be high; they were now even more rigid. "It takes two years to become a professional pilot," he insisted.

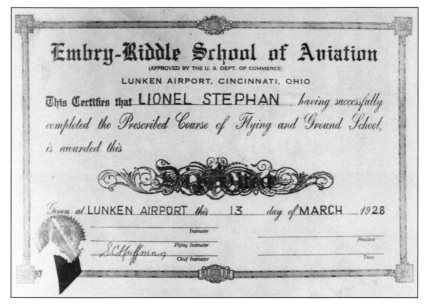

Embry-Riddle School of Aviation
(APPROVED BY THE U. S. DEPT. OF COMMERCE)

LUNKEN AIRPORT, CINCINNATI, OHIO

This Certifies that LIONEL STEPHAN *having successfully completed the Prescribed Course of Flying and Ground School, is awarded this*

Given at LUNKEN AIRPORT *this* 13 *day of* MARCH 1928

▲ One of the early graduation certificates from the Embry-Riddle School of Aviation.

Those two years were packed with instruction and safety awareness. The primary course included 10 hours of dual instruction, one hour of solo time, and 30 hours of ground school. Students in the advanced course flew another 50 hours of solo with frequent check rides and corrective counsel. Following those 50 hours, the candidate had to pass a written examination. If successful, he received a "limited commercial pilot's license." Only after adding another 200 hours of flying time to his log book could he earn a transport pilot license.

The Company made a distinction between teaching someone to fly and training someone to be a pilot. Instructors of the School maintained that anyone can "fly" after having a primary course and one hour of solo, but he is not yet a "pilot." Furthermore, 50 hours of solo flying and ground courses do not make fliers worth very much to industry. To be a pilot, claimed the School's instructors, the individual needs the primary and advanced courses, along with the ground courses and flying experience in many types of aircraft. He needs experience in airport and airline management, as well as hangar and maintenance work. "This way," said "Jiggs" Huffman, "the graduates of Embry-Riddle will be safe fliers and assets to the industry."

Mr. Riddle had another way of identifying the promising student. "I watched how a person treated the aircraft when he taxied to the ramp. If he was a 'hot dog,' I didn't want him. And I told him so, as soon as he climbed out of the plane."

Other schools apparently did not share Mr. Riddle's concerns. The lack of uniform quality of flying schools drew the ire of national figures such as Sen. Hiram Bingham of Connecticut, who in 1928 sponsored an amendment to the Air Commerce Act that

▼ The Embry-Riddle logo left no doubt as to its mission.

would require an annual examination and rating of civilian flying schools. He wrote:

> The Aeronautics Branch of the Department of Commerce daily is flooded with requests from people who ask: 'where can I obtain a good, reliable course in flying?' Owing to the lack of standardization, stabilization, or rating of the flying schools in existence today, this vitally important information cannot be given.

Others joined in urging the adoption of this amendment. The *Air Commerce Bulletin,* an official Department of Commerce publication, reported in 1929:

> The greatest number of accident causes are chargeable to personnel, and the greatest contributing factor in this category is poor technique on the part of the pilot. This, of course, reflects upon the flyer's original training and leads to the inevitable conclusion that still higher standards in flying schools are necessary.

The amendment was passed with ease. The terms of the amendment were to be administered by the Aeronautics Branch of the D.O.C. which was empowered to regulate equipment, personnel, and the size of the field. The Branch was also charged with controlling the number of students assigned and hours flown each day by instructors. Flight instructors and aircraft would have to be licensed.

Only five flying schools were endorsed by the Department on July 15, 1929. Embry-Riddle was one of them.

An Airline, Too

Mr. Embry and Mr. Riddle embodied a philosophy that would become a tradition with Embry-Riddle. Once something was accomplished, they refused to be satisfied with the status quo. Instead, they looked for new horizons. One of those bold, new adventures was the subject of an April 15, 1929, headline in the Cincinnati *Commercial Tribune:*

EMBRY-RIDDLE TO BECOME LINK IN GIANT AIRWAYS.

The founders of the School were determined to expand their operation, and they were not afraid to seek financial help from others. Embry and Riddle put up more than $10,000 of their own money, but more capital was needed to support the air mail services. Cincinnati businessmen contributed over $90,000, but this still was not enough.

Since the Company was a dealer for Fairchild Aircraft, Sherman Fairchild himself introduced Riddle to two highly respected New York investment banking firms—W. Averell Harriman & Company and Lehman Brothers. (Fairchild probably feared that the Company might sell other aircraft, such as the one offered by Glenn L. Curtiss, if he didn't help.) Fairchild acted quickly and with insight. He appealed to the investors for assistance. They agreed, and created a new financial enterprise called "Aviation Corporation of Delaware" on March 1, 1929.

Stock was underwritten by W. Averell Harriman & Company and Lehman Brothers and offered for public sale at $17.50 per share. Some of the more optimistic friends of Embry and Riddle envisioned as much as $1 million in sales. To the amazement of everyone, nearly two million shares were sold immediately, raising $35 million in cash.

Other aviation companies were added. The new organization was involved with aircraft manufacturing, engine development, sales, schools, and air passenger service.

On April 15, 1929, Embry-Riddle filed a charter of incorporation naming T. Higbee Embry as President and J. Paul Riddle as Vice President. By creating the corporation as a holding company, Embry-Riddle received the capital it needed. The Aviation Corporation of Delaware issued the stock, but the control of the operation remained with Embry and Riddle.

With the added cash, Embry and Riddle began a venture that was ahead of its time—even for them. They laid the groundwork for a unique concept called an "air university." The theory was that a primary school would train new recruits, while a "university" that specialized in advanced commercial aviation would provide professional aviators.

The new Embry-Riddle Aviation Corporation had earlier acquired Norton Field of Columbus, Ohio, and the Columbus Flying Service. Later it purchased the Portsmouth (Ohio) Flying Service. If these were successful, future plans included opening schools in other Ohio cities. The interest by potential enrollees was less than spectacular. Riddle attempted to gain more through advertisements and promotions, but nothing worked.

"Everybody was putting in schools everywhere," he said. "Our idea of an 'air university' never made it past the planning stage, I'm afraid."

The Embry-Riddle Company continued operating in Cincinnati as a subsidiary of the Aviation Corporation of Delaware, which was to become known as "AVCO." By the end of 1929, it controlled 81 different companies representing various aviation enterprises. Unfortunately, profits did not match its rapid growth.

The Corporation had lost $1.4 million. That was not much by today's standards, but in a depression-dominated economy it was simply more red ink than its management could tolerate.

Routes of the acquired air transport system did not connect effectively. "It was simply a mess," said Mr. Riddle. Often cut-throat competition for passengers between the mail and non-mail subsidiaries resulted in everyone losing money.

A new, super holding company had to be organized in 1930. This new subsidiary of the Aviation Corporation of Delaware was called "American Airways."

Embry-Riddle, then, was directly involved with the emergence of two of aviation's most powerful industries—AVCO and what today is known as "American Airlines." Yet in spite of the financial clout of these corporate giants, Embry-Riddle never took a back seat in corporate influence.

An Interlude

Embry and Riddle were still influential, respected authorities to whom the organization listened. Nonetheless, in terms of financial structure, the writing was on the wall. Reorganization was inevitable. Perhaps because they had survived when others had counted them out, Embry and Riddle, according to award-winning aviation writer Robert Serling in his book *Eagle—The Story of American Airlines,* "resisted a full takeover for a long time." Other historians claim that the tiny flight school stubbornly refused to sell their stock and attempted for nearly one full year to ride out this latest storm. According to the same writers, the Embry-Riddle Company reluctantly bowed to the obvious and sold its stock to American Airways in return for shares of common stock in the airline.

Mr. Riddle, however, put a different spin on the story. "We were running most of the operations of the airline part of the Aviation Corporation," he said. "We eventually sold our stock to the Aviation Corporation Airways in return for shares of common stock in the Corporation. It was simply a good deal for us all around."

Durward "Duke" Ledbetter, a pilot for American from 1928-1963 and a close personal friend of Paul Riddle, echoes Mr. Riddle's account.

As to which version is correct has been and will probably remain the subject of debate.

By agreement, Embry-Riddle maintained control of its own operating company. Its 270-mile route had become a part of the Aviation Corporation's aviation empire, providing it with an important midwestern link in the network of this growing airline.

Eventually, all the subsidiaries of American Airways were dissolved. Thus, one of the sadder headlines in the history of avia-

◄ **Loading an Embry-Riddle airplane with mailsacks was a daily ritual at Lunken Field in Cincinnati.**

tion appeared in the *Air Transportation Magazine* on September 30, 1930:

EMBRY-RIDDLE CLOSES DOWN OLDEST SCHOOL

The School's closing marked the end of a pioneering organization as well as the entrepreneurial partnership of T. Higbee Embry and J. Paul Riddle.

Mr. Embry moved to California, where the exuberant businessman lived until his death in 1946. Mr. Riddle went to New York, then to Dallas as a trouble-shooter in helping to structure the new organization. The corporate office was moved to St. Louis in March 1932, merging with the Universal Division to be called the "Central Division of American Airways." In less than one year, the energetic but cautious aviator J. Paul Riddle left St. Louis and the airline to seek his fortune in Florida. The man who had already become an aviation legend was only 32 years old.

The Embry-Riddle Company's association with Lunken Airport and its love affair with Cincinnati was now history. An era had passed; another was yet to be born.

3

World War II

*"The growth and progress of the
Riddle Technical School here
as well as its importance in the national
defense setup is remarkable."*

— MIAMI DAILY NEWS, 1941

John Paul Riddle didn't move to Miami in order to retire or to bask in the sunshine. He saw a rich future for aviation in Florida. He was convinced, in fact, that Miami could develop into a major gateway to South America.

Shortly after he established permanent residence in Florida, Riddle formed the Palm Beach Aero Corporation. This latest venture had a modest beginning, but gained support from some local influential aviation enthusiasts such as Marshall "Doc" Rinker, Jake Boyd, and Paul Butler. In addition, Mr. Riddle formed the Miami Aero Corporation and the Florida Aero Corporation. All three served as foundations for his future endeavors.

In October 1939, Riddle formed an equal partnership with John G. McKay, a prominent Florida attorney and a former Army Air Corps lieutenant. Together they established a seaplane operation which they named "The Embry-Riddle School of Aviation" along what is now MacArthur Causeway. Shortly thereafter, the School added another training center at Miami's Municipal Airport to train students in consortium with the University of Miami.

As a basis for further expansion, the School created "divisions" for various operations. Each had its own general manager, while Riddle served as president and McKay as vice president and legal advisor.

Overseas, the sounds of war grew more intense. Hitler's armies goose-stepped into Poland. Suddenly, England became vulnerable. Across the Pacific, Japan was flexing its muscles, threatening to rule the Eastern world.

Two years before America's direct involvement with the war, Riddle (now fondly referred to as "Boss Riddle") and McKay prepared the School to train pilots and mechanics. Both men sided with the outspoken Brig. Gen. William (Billy) Mitchell, who predicted in 1925 that air power would play an important role if America ever signed a declaration of war.

On Sunday morning, December 7, 1941, the entire nation knew just how right they were.

Cadet Training

The United States needed to train pilots and mechanics in a hurry, and turned to those who could help. The Embry-Riddle School was ready and offered the services of its 390 employees, including 87 flight instructors.

The Army Air Corps entrusted Embry-Riddle along with other private schools with the primary training of cadets while the Corps devoted its own fields to advanced training.

Britain's Royal Air Force, too, sought schools for its future pilots, and Embry-Riddle was awarded a contract to provide this training.

Paul Riddle and John McKay had already expanded their training sites to four satellite airfields.

Carlstrom Field in Arcadia, Florida, was the largest under the Southeastern Air Corps Training Center of Maxwell Field, Alabama. Unlike other military bases, this 1,300-acre, flat, dry field was no make-shift operation. It had a look of permanency.

◀ **Cadets trained by Embry-Riddle prepare for flight at Carlstrom Field.**

▲ Carlstrom Field, near Arcadia, Florida. Formerly a major flying field for the U.S. Army, Carlstrom became the largest civilian training field and was operated by Embry-Riddle until its closing in 1945. Today it is the site of the G. Pierce Wood Memorial Hospital.

Hangars formed its outer rim; inside were barracks, an infirmary, mess hall, classrooms, recreational facilities, even the legendary swimming pool into which a cadet was thrown by his classmates upon completing his first solo flight.

Leonard J. Povey served as General Manager over the fast-paced program that provided instruction for 500 cadets at one time over a nine-week course that included 60 hours of flight.

Both American and British pilots were trained at Carlstrom. When the first wave of RAF cadets arrived, the Associated Press reported:

> The eyes of 99 Britons lighted when they glimpsed the new $300,000 Riddle Aeronautical Institute, with its lines of

▶ The first cadet to report in at Carlstrom was Frank Beeson, being greeted by Len Povey (far right), as a second cadet and another company official look on. Lieutenant Beeson was killed in action over New Guinea in July 1942.

trainer planes, modern barracks, swimming pool, and tennis courts. "The Royal Air Force will turn green with envy when it hears about this," exulted one cadet as he examined his quarters.

During World War II, one of the widely published demonstrations of patriotism was the result of Carlstrom's staff. In May 1943, *The American Pilot and Aircraftsman Magazine* published "Carlstrom Field's Twelve Ways to Aid Hitler":

1. Griping about food.

2. Taking ground school lightly, thinking if you can fly a plane that is all you need.

3. Ground-looping through carelessness, resulting in time lost in plane repair.

4. Goldbricking in physical training, passing up the opportunity to develop strength, endurance, ability and recreation.

5. Reporting to sick call with acute "drillitis" to evade duties that instill personal discipline so important to the fighting man.

6. Using open post to undo physical benefits of training.

7. Failure to observe general rules of military courtesy. Save the chip on the shoulder for Adolf.

8. Poor use of time when you might be writing home. Your folks want to know how you are and their spirits must be kept up as much as yours.

9. Giving up if you are disqualified for further flight training. Remember, chickens don't fly, but they are darned useful birds.

10. Letting foresight sag. You will have your chance to fight, but you have to make good here first.

▶ **Aircraft Overhaul Department at Carlstrom Field. One of three locations of the Aircraft and Engine Division, Carlstrom Field was kept busy. The other two locations were in Miami and Coral Gables. The division overhauled more than 3,000 engines, 700 aircraft, and 21,000 aeronautical instruments of the nation's war effort.**

▲ Riddle Field, south of Lake Okeechobee in Florida. Here the company trained RAF pursuit and bomber pilots in primary and advanced flying from 1941 to 1945. The field is now the Airglades Airport.

11. Losing your faith in worship. An hour in church might give you a fresher outlook for the week.

12. Forgetting there are several million more just like us in this war.

Riddle Field, a 2,500-acre site located at Clewiston, Florida, was the site of the No. 5 British Flying Training School and could accommodate 250 cadets. Used primarily for RAF training, it provided primary, basic, and advanced phases of flight instruction.

Operating as the "Riddle-McKay Aero College" Division, Embry-Riddle provided complete training for bomber and pursuit planes, including combat maneuvers, under the watchful eyes of General Manager G. Willis Tyson.

Dorr Field, like Carlstrom—also operating under the "Riddle Aeronautical Institute" Division—was located near Arcadia. Flight training began here before the end of summer 1941; by the new year, cadets moved into the dormitories. Embry-Riddle Field, near

Union City, Tennessee, contained 870 acres, and was assigned to the "Riddle-McKay Aero Institute of Tennessee" Division. H. Roscoe Brinton, the first General Manager, initially guided the primary flight training for 100 cadets of the Army Air Corps.

The Seaplane Base in Miami became increasingly important, as it was one of the few places where Navy pilots could receive the basics in this specialized type of training. Among the students who earned their seaplane ratings here was one young officer

▲ **Dorr Field, near Arcadia, Florida. Many Army Air Corps cadets began their training here, a site that is now a maximum security prison.**

from Huntsville, Tennessee, who later distinguished himself in politics as a United States Senator—Howard Baker, Jr.

Prior to the attack on Pearl Harbor, the Miami Landplane Base at Chapman Field located south of Miami on Biscayne Bay trained young men from both the civilian ranks and from the Civil Pilot Training Program [CPTP]—a project of the Civil Aeronautics Administration [CAA] to prepare college students as pilot reserves in the event of war. "One of our best instructors there was Tine Davis," said Mr. Riddle, "who could fly anything with wings." Mr. Davis, who later became one of the founders of the popular Winn-Dixie stores, was a benefactor of Embry-Riddle and later served as a member of its Board of Trustees.

In the spring of 1940, one person changed the image that flying was a "for men only" career. Dorothy Ashe, daughter of the University of Miami's president, B. F. Ashe, was Embry-Riddle's first woman pilot to solo under the CPTP. Others followed. In less than a year, the School hired Helen Cavis—one of the 160 women in the United States who held a commercial certificate—as its first woman flight instructor.

Miss Cavis was originally hired not only to teach women pilots, but men as well.

Not everyone applauded this turn of events. Some had a hard time accepting the idea of a woman working in the previously male-dominated environment of aviation. One of the more brazen cadets was even bold enough to approach Mr. Riddle in his office, complaining that "no women—'schoolmarms,'" as he and some of his companions called them—"should pilot an airplane."

"Bah!" retorted Riddle. "Any fool realizes that a plane doesn't care who's at the controls. Besides, in my opinion she flies a lot better than most of the men here...including you." Tilting his head and raising one bushy eyebrow, Riddle asked: "Is that what's really bothering you, young man?" The red-faced cadet made an abrupt "about face" and left without saying another word.

The women must have done something right. By the end of 1942, the Seaplane Base became an all-women operation, run and staffed by women headed by Ruth Norton, its director.

▶ RAF graduation at Riddle Field. Pictured is the cadet formation for the graduation of pilots trained for the Royal Air Force by Embry-Riddle instructors. More than 2,000 RAF cadets stood proudly as they won their wings at Riddle Field. Biplanes stand to the rear of the formation, and advanced trainers are in the distance.

Most of the pilots who graduated from the Embry-Riddle School soon flew the skies above territories with strange-sounding names, such as Saipan, Leipzig, and Iwo Jima. And many never returned.

Technical Division

As the Government called for more pilots and airplanes, Riddle and McKay realized that mechanics and technicians would be needed to "keep 'em flying." Consequently, in November 1940 they leased the south wing of the "Old Fritz Hotel" on N.W. 27th Avenue and 32nd Street in Miami for ground school and technical instruction.

The hotel, dubbed by many as one of Miami's "white elephants," was built in 1925 as a sanitarian health hotel by millionaire Joachim Fritz. With the stock market crash on Wall Street in October 1929, many of its clientele no longer could afford this luxury. The five-story structure was forced to close its doors in 1930. It sat dormant for several years, until a poultryman decided to use it as a chicken hatchery. When Riddle and McKay selected the building as the School's technical training site, the students

▼ The Seaplane Base in Miami was a training area for many military pilots during the war.

▲ Both the United States and British flags flew over Embry-Riddle training fields during the war.

and faculty referred to it as the "Aviation Building." The citizens of Miami, however, had baptized it with a somewhat less complimentary title—"The One Million Dollar Chicken Coop."

Yet Riddle and McKay proved they could even transform a "white elephant chicken coop" into a useful entity.

The August 17, 1941, edition of the *Miami News* agreed:

> The growth and progress of the Riddle Technical School here as well as its importance in the national defense setup is remarkable. Its approximately 500 students from North and South America are studying all phases of aircraft, engineering, construction, flight operation, and maintenance. More than 1,000 skilled craftsmen, men and boys with little or no previous aircraft experience, will be turned out here this year to join scores of other graduates working in aircraft plants all over the country.

The Technical Division operated seven days and five evenings a week. In 1942, the Army sent more trainees. The School tripled its training schedule, and classes were taught 24 hours a day, seven days a week. The concentrated program of study produced trained technicians, sometimes after only six months of classes.

As rapidly as students completed their courses of study, there were still not enough men to fill all the vacancies in the work force. To fill these gaps, the School opened its technical courses to female students in 1942. The women were taught and performed all types of work within the limitations of their physical strength.

Each student learned even more by "doing" than they would through textbooks, alone. The highly individualized method of training included a personal instructor who guided the student step-by-step through every phase of maintenance.

Unique demands required unique training methods. In some instances, students had to learn to work by touch alone, for on the fighting fronts they would sometimes have to work in total darkness.

There were no semesters, no summer vacations, nor a lot of the other things normally associated with today's institutions of higher education. Only when the instructor cleared him, and the student successfully passed his exams, was he ready to work at an airplane factory or hangar.

One of those aircraft plants was a part of Embry-Riddle's Overhaul Division created in July 1942 on 20th Street in Miami, a short drive from the Aviation Building. *The Fly Paper*—the official in-house publication of Embry-Riddle—described the fast-paced activity:

> In a hangar at the back of the Aviation Building, engines for overhaul are received at one end of the building . . . pass through a production line procedure . . . and, a few hours

▶ The Aviation Building. The Technical Division occupied the entire former hotel from 1940 to 1944, and the school returned from 1952 to 1965. It was known as the "$1 Million Chicken Coop," named for its use before 1940. The building was demolished in May 1971 to make room for a juvenile court and youth hall.

later, pop out of the other end of the building completely overhauled and ready for service, including even the required "run in" in the engine test house.

The major problems confronting this new division included obtaining tools, equipment, and personnel. Tools and equipment often had to be designed and manufactured by the instructors; women provided the solution to the personnel problems. In fact, by June 1943, women constituted more than one-third of the division's employees.

The School also set up an Instrument Department in suburban Coral Gables in which technicians repaired and refinished all types of aircraft instruments.

Since many of the potential students were from South America, Riddle and McKay offered dual-language instruction. As a result, the School created a partial translation of aircraft nomenclature into Spanish and Portuguese—one of the first such authoritative works ever attempted.

A contract was awarded to Embry-Riddle through the CAA to participate in the Inter-American Aviation Training Program.

The first 70 students from Latin America arrived in January. In just seven months, 21 students from nine countries completed their courses of study.

Embry-Riddle trained men and women throughout the course of World War II, until August 6, 1945, when a 29-year-old Air Corps Colonel, Paul Tibbets, flew the *Enola Gay* over Hiroshima and gave the order to drop an experimental bomb that not only ended the War but also changed the course of history.

Transition

By the time General Douglas MacArthur signed the peace treaty on September 2 aboard the *U.S.S. Missouri,* more than 26,000 men and women had earned their wings or technical licenses through Embry-Riddle. Through all of this, the School's remarkable record of safety won the praises of every branch of the Armed Services.

In addition, the Aircraft and Engine Division overhauled more than 3,000 engines and 700 aircraft. The Instrument Department in Coral Gables repaired and refinished at least 21,000 aeronautical instruments.

For the war-time programs, the School successfully trained students for the fighting fronts. It was only fitting that it would now prepare to equip students equally well for a world at peace.

4

Toward a University

*"Be it resolved that J.R. Hunt be,
and he hereby is, elected president of
Embry-Riddle Aeronautical Institute,
at a salary of $900 per month, to serve
at the pleasure of the Board."*

— MINUTES FROM THE BOARD OF TRUSTEES,
SEPTEMBER 3, 1963

Changing of the Guard

"It was the hardest decision in my life, but it's something I had to do," said John Paul Riddle when he sold his interest in the School to John McKay in September 1944. Over the past 18 years, Mr. Riddle had carved a niche for himself in the history of aviation.

Most men, we suppose, would be satisfied. But John Paul Riddle was not like most men. He was not running away from a challenge; he was setting his sights on a new one. The *Fly Paper* put the story into a nutshell:

> John Paul Riddle is leaving now for ventures in a new world, the world of aviation fast developing in neighboring Brazil. There he found the same sky, the same air, but new horizons, far reaching and, as yet, unconquered. There are new dreams to come true.

Those new dreams included the "Escola Tecnica de Aviacao" [Technical School of Aviation] in Sao Paulo, which began as a cooperative venture involving Embry-Riddle and the Brazilian Air Ministry.

The new school in Brazil was the result of a suggestion by Mr. Peter Ordway, a graduate of Embry-Riddle and former Dean of

▲ John G. McKay, Jr. formed a partnership with J. Paul Riddle in 1939 to reopen the Embry-Riddle flight school in Miami. He became the sole owner in 1944 and died in 1951; by then the company had become much more than a flying school.

Students and Advertising Director. "Peter Ordway was so supportive of Embry-Riddle that when his son was born, he 'pre-enrolled' the infant for the freshman class eighteen years later," said Mr. Riddle. Heading the operations in Brazil were David Beatty, who would later gain fame as a backer for several motion pictures at M.G.M., and William Lehman, a former congressman, a long-time friend and former instructor for Mr. Riddle.

After Riddle dissolved his partnership with McKay, the Brazilian school was no longer a part of Embry-Riddle. Riddle remained with it, nurtured it to prosperity, then turned it over to the Brazilian Government. In just three years, the institution graduated 3,500 students; at one time, it employed 650 North American instructors.

▼ Early recruitment efforts. The Embry-Riddle School of Aviation searched for students interested in civil aviation.

Back in Miami, Embry-Riddle was beginning a series of transformations, including two name changes and three moves.

The "Embry-Riddle International School of Aviation," as it was now called, consisted of the Technical School in Coral Gables; the Flight Division, with two bases at MacArthur Causeway and Chapman Field, now devoted to training pilots for commercial and business aviation; and the Aircraft Division at N.W. 20th Street and Chapman Field.

The military no longer needed private schools to teach its pilots and mechanics. The new market was civilian training, although the School signed contracts with the Veterans Administration to instruct those qualified under the G.I. Bill of Rights and with the Free French Republic to provide advanced flight training for its naval cadets.

John McKay became the new president of the School, appointed new officers, and laid the groundwork for the new challenges of the postwar era.

▲ **Piper Cubs and other trainers line the ramp at the Opa-Locka Airport. The Embry-Riddle International School of Aviation, the company's name at the time, was consolidated here in the northwestern Miami region from 1947 to 1952.**

New Settings

All divisions of Embry-Riddle moved to the Opa-Locka Airport in October 1947. The airport, built by Glenn Curtiss and enlarged by the Navy during World II, provided all the spaces necessary

▼ Upon the death of her husband in 1951, Mrs. Isabel McKay became the president and general manager of Embry-Riddle. She reorganized the company in 1961 as a non-profit school, appointed Jack Hunt her successor in 1963, and remained as a member of the Board until her death in 1972.

for the Flight and Technical Schools. A large Navy-constructed hangar, an adjacent hangar, and a utility building were used by the School. The approximately 400 students who enrolled over the next few years utilized the dormitories and cafeteria at the airport.

During the Korean War, the School was once again called upon by the military to train mechanics. The nation's newest branch of service, the United States Air Force, contracted with Embry-Riddle to train airmen in the fundamentals of airplane maintenance, from the basic instructions on how to care for hand tools to the complex principles of aerodynamics, electronics, and hydraulics. More than 1,500 Air Force recruits trained at Opa-Locka during a 16-month period.

The geography was different, but the tradition continued. Embry-Riddle insisted on rigid adherence to safety standards. "The School enjoys a high safety record," reported the *Air Trails Magazine* of November 1948. "Correct flying habits and a sound foundation on the ground to sharpen efficiency in the air is a 'must' for the beginner or the commercial pilot seeking additional ratings."

John McKay ran the School, making most of the major decisions until he suddenly died in 1951. Mrs. Isabel McKay, his widow, who had served as his assistant since 1945, assumed the

▲ Tamiami Airport in Miami. Embry-Riddle flight training occurred here from 1952 to 1965 while technical training was located eight miles away in the Aviation Building.

presidency, becoming the only woman to head an aviation school in the United States.

For one so wrapped-up in all phases of aviation, it might be assumed that Mrs. McKay was an avid pilot. "Not really," she confessed to a reporter for the *Miami Herald*. "I started taking flying lessons once, but gave it up after only four hours in the air."

In 1952 the Marine Corps occupied the Opa-Locka Airport and forced the School to relocate once again. This time the officers, classrooms, and shops were moved back to the "Aviation Building" (still referred to as the "Chicken Coop" by most people), while the Flight School operated out of the Tamiami Airport, eight miles away.

While at Tamiami, the School renewed its consortium with the University of Miami. The two schools combined to provide students with the necessary requirements for a four-year degree in Aviation Administration and a two-year certificate in the Business Pilot Program. Four years later, Embry-Riddle provided flight training for the University's Air Force ROTC students.

Non-Profit Status

Embry-Riddle's enrollment increased steadily over the next few years because of the lure of aviation for students not only in America, but throughout the world. The School's enrollment read like a roster at the United Nations. There were young men from China, Turkey, Lebanon, Israel, Egypt, Pakistan, Norway, Belgium, Switzerland, England, Bermuda, Cuba, Trinidad, Canada, Columbia, Bolivia, Honduras, Venezuela, Guatemala, and Panama.

It was this world-wide appeal, in fact, that enabled Embry-Riddle to continue operating. Despite all efforts to promote the Institute within the Miami community, the School was better known outside its own hometown. By 1958, its 1,000 students represented 44 states and 21 foreign countries.

Aviation had reached the jet age. If Embry-Riddle was going to serve the industry, it had to adjust. "We are keeping pace with aviation developments by including a study in jet theory, and we will offer courses in jet engines in the near future," promised Mrs. McKay.

The School, now named "Embry-Riddle Aeronautical Institute," offered courses for jet specialists and electronics technicians. With this more sophisticated curriculum, there were, of course, greater expenses for the day-to-day operation.

Until now, nearly every dollar of Embry-Riddle's income came directly from student tuition. If the Institute were to survive, other sources of support had to be generated. One way was to gain acceptance as an independent, non-profit institution.

After more than a year of lobbying through persuasive rhetoric, Mrs. McKay, Paul Freedland, and other Board members were successful in winning approval from the State of Florida to grant non-profit status to the Institute. On September 1, 1961, it could receive financial assistance from philanthropic organizations and, if necessary, even from the Government.

New Faces

Mrs. McKay continued serving as president until she was certain that the Institute would function smoothly. By New Year's Day, 1962, she demonstrated that confidence when she turned over the reins to G. Ross Henninger. Mrs. McKay became Chairman of the Board of Trustees.

Mrs. McKay's faith in Mr. Henninger's ability to lead the Institute may have been misguided. For a variety of reasons, the arrangement simply did not work out. Mrs. McKay knew it; the Board knew it. On June 4 of that year Mrs. McKay called for a special meeting of the Board. "This is a sad day for me and for Embry-Riddle," she said. "I must ask that you, the members of the Board, remove Mr. Henninger as president."

Mr. Henninger, however, protested this motion, and appealed

to the Board for support. After heated debate, the Board voted unanimously either to accept the resignation of President Henninger, or, if he refused to resign, to terminate him.

Mr. Henninger resigned.

A long-time friend of the Institute, Dr. Grover Noetzel, accepted the role as interim president while the Board searched for a new leader.

On September 3, 1963, the Board announced that it had found that leader.

Mrs. McKay read the official pronouncement: "Be it resolved that J. R. Hunt be, and he hereby is, elected president of Embry-Riddle Aeronautical Institute at a salary of $900 per month to serve at the pleasure of the Board."

Jack Reed Hunt had worked with the Institute since June of that year as a consultant. Hunt, a dashing former Navy commander, had gained international fame as the pilot of the first non-stop, round-trip flight across the Atlantic in the U.S. Navy blimp ZPG-2. Hunt led the 14-man crew of the airship Snowbird for 364.2 hours over 9,448 miles from March 4 to 16, 1957. The route took them from Massachusetts to Portugal, to Africa, to Puerto Rico, to Cuba, to Florida. It was the longest non-refueled sustained flight ever made.

For this amazing accomplishment, Fleet Admiral William F. "Bull" Halsey awarded Commander Hunt the Distinguished Flying Cross. President Dwight D. Eisenhower presented him with the Harmon Trophy at a special ceremony in the Oval Office of the White House.

Soon after he was discharged from the service, Hunt established a solid reputation as a real estate investor. One of the reasons for his success was his unusual ability to recognize the value of certain properties that escaped others.

Jack Hunt didn't have to wait long for an opportunity to prove what sort of leader he would be. Within a few weeks, the Dade

County Port Authority announced that the Tamiami Airport would be phased out of existence. Embry-Riddle had to find a new home, and do so quickly.

Selecting a Site

Transporting a school can produce some nearly insurmountable problems. The most obvious is the physical transporting of equipment, books, and machinery from one location to another. Add to these the fact that the holdings included several aircraft and the problem becomes even greater.

Other questions had to be answered as well. What effect would the move have on recruiting students? Would this affect the retention of the faculty and staff? How would accrediting agencies view the move? "Embry-Riddle is virtually unknown in Dade County," said President Hunt to the Board in January 1964, "yet, internationally, the School enjoys a fine reputation."

Hunt and the Board developed a plan of attack. Instead of fighting the apathy of Miami, it would move the Institute elsewhere.

Twenty-two sites were visited by officials. Each one offered certain advantages. However, holding the most promise were five Florida locations: Sebring, Fort Pierce, Plant City, Dade County (outside of Miami) and Daytona Beach.

Embry-Riddle suddenly found itself in a new role. All five communities really wanted the Institute. Daytona Beach fought harder than the others. Led by Gary Cunningham, a bold entrepreneur who had gained a reputation for getting what he went after, the Daytona Beach "Committee of 100"—a group of citizens dedicated to invite new business into the area—pledged its support not only to roll out the welcome mat, but aid in the actual moving of the campus 250 miles north.

▶ **President Dwight D. Eisenhower** *(left)* **presents Commander Jack Hunt** *(right)* **with the Harmon Trophy at a special ceremony in the Oval Office of the White House, as Fleet Admiral "Bull" Halsey looks on. Commander Hunt also earned the Distinguished Flying Cross.**

▲ Jack Reed Hunt was the president of Embry-Riddle Aeronautical University from September 1963 until his death in January 1984.

▲ "Operation Bookstrap." Thirty-four heavily loaded trucks made a convoy for the return drive to Daytona Beach early on the second day of the operation. More than fifty volunteers went along to help.

"Operation Bootstrap" to Daytona Beach

"This is a victory for the good image of the Daytona Beach area," reported the *Daytona Beach Evening News* on April 27, 1965.

Moving a family or a small business from one locale to another is expensive; moving an entire institute of this magnitude could be financially prohibitive. Were it not for organizations such as the Daytona Beach Jaycees, the "Committee of 100," and local residents willing to donate time and energy into transporting everything north, the move to the Daytona Beach area could never have been accomplished.

Underwriting the costs of the move was no small challenge. The School's accountant put a sharp pencil to paper and determined that it needed $75,000 before the first library book could be packed.

▲ Jack Hunt *(left)* helps to load one of the trucks bound for Daytona Beach.

Vice president of the "Committee of 100" was Philip H. Elliott, Jr., a prominent attorney of Daytona Beach, who later would be selected as the School's legal representative and member of its Board of Trustees. He was one of the first to support the project.

But his faith was not shared by everyone. Elliott recalls: "Most of the community leaders showed little enthusiasm for the move. Some bluntly asked, 'How can they be an asset to the area if they have to borrow money just to move in?'" Gary Cunningham turned to a young, aggressive insurance executive, John C. "Jay" Adams, to serve as chairman for the drive. Together, they knocked on doors of local business owners and interested citizens. "This will be a loan, not a gift," both Cunningham and Adams promised.

Whether it was the community's confidence in the Institute or respect for Cunningham and Adams that determined the response is debatable. Nevertheless, every dime of the $75,000 was raised in just four days. A convoy of trucks and drivers was organized. On April 24, a total of 56 volunteers met at 7:30 a.m. at the Daytona Beach Airport, climbed aboard 34 donated trucks and vans, and drove south on U.S. Route 1 toward Miami.

Throughout that afternoon and evening, the crew loaded each book and piece of equipment belonging to the Institute aboard the vehicles. By 3:00 a.m., the final rope was secured. Most of the volunteers slept in the trucks until sunrise.

Slowly the vans pulled away. Reporter Bob Desiderio called it "Operation Bootstrap" and described the caravan in the *Daytona Beach Evening News:*

> The trucks were a sight to behold. There were little ones, big ones, soft drink trucks and carryalls.. . . It looked like a scene out of *Tobacco Road* or *The Grapes of Wrath.*

After nine hours, the convoy chugged into the Daytona Beach Airport only 30 minutes behind schedule to a welcoming party unlike that ever imagined by the exhausted crew.

"It was a royal welcome," remembers Cunningham. "Scores of

Jaycees and others cheered our caravan as we drove up Midway Road. We were tired, hot, sweating and unkempt. We must have looked like bums, but they greeted us like conquering heroes. When I got out of the truck, one of them handed me a cold beer. That was the best damned beer I've ever had in my life."

When the last empty bottle was tossed in the trash can, no one could deny that "Operation Bootstrap" was an unqualified success. The entire operation—lock, stock, and airplane—was transported without a scratch from Miami to Daytona Beach.

Original plans called for Embry-Riddle to relocate to the Ormond Beach Municipal Airport about 10 miles north of the heart of Daytona Beach. As a temporary set-up, however, Jack Hunt and his staff leased temporary quarters at the Daytona Beach Airport and planned to operate there, while the permanent campus was constructed in Ormond Beach along the Tomoka River. However, rumblings of protest came from a significant number of residents in Ormond Beach who campaigned against the establishment of Embry-Riddle at its airport. It seemed as though the 10-mile move to Ormond Beach would be more difficult than was the 250- mile caravan just made from Miami.

"Those little airplanes could crash on one of our churches and schools at any minute," shouted one resident during a city council meeting.

"For some reason or another, the people always expected airplanes to crash into schools and churches," remembers Gary Cunningham. "The pressure became too great, and we could never overcome that mind-set."

The School concluded operations in Miami in April. One month later, it opened its doors for classes in Daytona Beach. Its total assets were a half million dollars' worth of equipment, a determined administration, and the whole-hearted support of a

◄ A volunteer from the Jaycees, one of 56 people who volunteered to help with the move, prepares to start the journey from Miami to Daytona Beach.

progressive community.

During its nearly 40 years of operation in Miami, Embry-Riddle had trained nearly 40,000 students. An impressive number, perhaps, but it was only a prelude to what would happen in the next three decades.

5

Growing Pains

*"With all this building
came unexpected expenses.
We were literally at the bottom
of the financial barrel."*

— JACK R. HUNT, 1966

Building Boom

The Embry-Riddle Aeronautical Institute opened its doors for classes in Daytona Beach on May 19, 1965, (President Hunt's and Mr. Riddle's birthday) 39 years after the Embry-Riddle Flying School offered that first flying lesson in Cincinnati.

The Institute was divided into three divisions—the College of Engineering, the Airframe and Powerplant Technician School, and the Flight School. Enrollment that summer term numbered 239, including the 168 students who relocated from Miami.

The Institute operated at the Daytona Beach Municipal Airport in five leased buildings. The "Naval Reserve Building" contained the College of Engineering on the upper floor, and part of the Airframe and Powerplant School on the ground floor. Administrative offices were located in the "Army Reserve Building." A third building off Midway Avenue housed more of the Airframe and Powerplant School. The dormitory was half of a city-owned building shared with the Kansas City Athletics Baseball Club.

The Flight School and remainder of the Airframe and Powerplant School were at a hangar leased from the Daytona Beach Aviation, Inc. The 9,600-square-foot hangar had been hauled from the Tamiami Airport by Daytona Beach Aviation, which added another 5,760 square feet to it.

"With all this building came unexpected expenses. We were literally at the bottom of the financial barrel," recalled President Hunt. "There was only one thing to do if we hoped to survive. Five trustees and myself signed personal notes for five thousand dollars so that the next payroll could be made."

Their faith paid off. More than 500 students registered for fall classes.

"Now our problem was not can we grow, but how fast can we afford to grow," said President Hunt with a sigh of relief.

The increasing enrollment proved both a blessing and a curse. One year later, when the student body reached 750, the School had more students than it could handle with existing facilities. An additional 20 airplanes were leased. After unsuccessfully attempting to persuade the City to relocate the Kansas City Athletics, the School used four local hotels to house 300 students.

Buildings had to be added in order to keep up with the influx of students. In February, two prefabricated buildings were set up near the Naval Reserve Building and were used as a library and a combination student center/book-store. The City leased to Embry-Riddle a 1.5-acre site west of the terminal on which the School constructed a hangar large enough to include four much-needed classrooms.

Enrollment reached 1,015 students the next fall. More dormitory space was needed promptly. The School leased a new apartment complex (later named the "Alhambra Apartments") on Nova Road for 287 students, and purchased the Mount Vernon Motor Lodge on South Ridgewood Avenue for 40 more students.

Progress toward a permanent campus in Ormond Beach had come to a screeching halt. Along with growing citizen fears of "those little airplanes that could fall out of the sky," disagree-

▶ Shown here is the flight line at the Daytona Beach Municipal Airport in 1966. The hangar in left-center had been hauled from the Tamiami Airport by another company.

ments arose between Embry-Riddle and the city officials as to how large the School could grow.

Jack Hunt had many virtues, but patience was not one of them. "It was either put up or shut up," he said. "Ormond Beach shut up, so we simply quit wasting each other's time."

One month later, Embry-Riddle had its permanent home. The school arranged with the city fathers of Daytona Beach for use of an 85-acre tract in the northeast section of the Airport. Embry-Riddle agreed to purchase the property for $42,500 per acre, plus pay $50,000 for water and sewerage preparation. In addition, it was allowed to purchase the land piecemeal over the next three years.

"There was no question about it. Daytona Beach wanted us to stay where we were," said President Hunt. The next few months were saturated with building and phenomenal growth.

The first shovel of dirt was turned on October 17, 1967, for a new 386-student dormitory. Every penny of the $75,000 loaned by local residents toward the move was repaid within only 19 months. The faculty and staff increased from 38 to 180. The annual payroll was $860,000, and the economic impact on the community surpassed $2,500,000. More buildings were financed by federal loans, federal grants, private donations, and school income. One of the more "radical" designs was named the new "Academic Complex." The original drawings proposed rectangular buildings.

"Not flexible enough," insisted President Hunt. "Conventional buildings waste too much gawd-dang (one of his favorite expressions) space." He went on to explain that a rectangular building loses up to 25 percent to corridors, stairways, and storage areas. A hexagonal design, on the other hand, loses only 10 percent. Hunt's ability to persuade was demonstrated when three hexagonal buildings connected by covered walks were erected east of the airport.

The buildings didn't appear by magic, although some might consider it a miracle that Embry-Riddle was able to raise the necessary $1.3 million. Once again, "Jay" Adams was asked to spearhead a local campaign for $325,000—a necessary requirement before any matching or government funds were awarded. In order to meet this quota, Adams had to put his signature on a personally-secured promissory note in order to generate the needed cash.

Because of the ever-increasing student enrollment, classes were actually conducted in the new "Academic Complex" in February 1969, although the building was not completed until the fall.

The Flight Line reflected the fact that the Institute was growing faster than anticipated. It was a mixed breed of types and color schemes, including Cessna 150s, Piper Cherokees, single-engine Aero Commanders, three classic twin-engine Beech B-18s, plus a blue and yellow Stearman biplane rebuilt by some of the students.

In 1968, Embry-Riddle reorganized, and purchased eight acres of land. The groundwork was laid for recognition as a major institution of higher education. That same year, it received accreditation from the Southern Association of Schools and Colleges (SACS).

The granting of accreditation did not come easy. Jack Hunt appeared before the SACS assembly of delegates in December 1968 with a mission to convince them that Embry-Riddle was more than a glorified flight school. He introduced to the assembly his concept of a "University of the Air." He explained that Embry-Riddle was pioneering a new concept in flight training which involved a second student pilot riding along in the back seat of a training flight as an observer.

The unique training system was called the "1 + 1 Gemini-flight Technique." Hunt explained: "When a student is operating the controls, he often tends to worry more about what the airplane is doing and not what he is doing. Sitting in the back seat, however, he can learn from the mistakes being made by the student pilot at the controls. The next time they go up, they swap places, and the other student learns from his mistakes."

Hunt's argument was convincing primarily because he believed in what he was saying. On the many occasions he hosted a visitor on the campus, he pointed to a Piper Cherokee and exclaimed: "This is a classroom; it just happens to look like an airplane."

Those who were at the SACS gathering testify that it was Jack Hunt's pride in Embry-Riddle's product that was most persuasive in winning approval for accreditation.

Financial woes again dominated the concerns of the Administration. "We were simply running out of cash and credit," says "Jay" Adams. Adams, who now served on the Board of Trustees and had earned a reputation in the community as one of the more vocal supporters of the School, tried in vain to get loans from area banks during the early fall of 1969. Not only was the school obligated to pay its debts for its rapid building program, but, unless it could get a quick infusion of cash, it could not meet the December payrolls for faculty and staff.

Jack Hunt reached into his bag of imaginative solutions and produced yet another unique idea. On November 15 he announced that any student who prepaid his next semester's tuition before the end of the current term could return to school on January 6; otherwise, he would have to report on January 2. Hunt hoped the extra four days vacation would be enough of an incentive to gather enough income to meet the immediate needs.

The plan worked. Faculty and staff were paid on time, and the School could get on with the business of education.

Progress came in the form not only of academic standing and building construction, but in other areas as well.

Until 1967, Embry-Riddle, like the rest of the aviation community, was dominated by men. Jack Hunt determined it was time to open the doors to female flying buffs. In the summer of '67, three

◄ **A dormitory for Daytona. Dorm I, the first permanent structure of Embry-Riddle in Daytona Beach, was built in 1968 to house 386 students.**

co-eds—Penny Choranec, Linda Larsen, and Jackie Wimpy—formed the vanguard of what promised to be a rapidly expanding group once the word got around.

"There's a big future for women in aviation," predicted Hunt. "Of course, it's going to take a little longer to get them into the cockpit of a commercial airliner. It takes time for ideas to change."

The ideas eventually changed about a lot of things, including the academic status of the Institute.

Graduating to a University

The three divisions were dissolved, replaced by two colleges: The College of Aeronautical Studies and the College of Aviation Technology. The School's engineering technology program had already been accredited by the Engineering Council for Professional Development; the flight program and airframe powerplant program had earned approval of the Federal Aviation Administration (FAA).

On June 9, 1970, the blue and gold sign at the entrance to the campus that welcomed visitors to Embry-Riddle Aeronautical Institute was repainted. In place of the word "Institute," was an even more impressive word—"University."

The impact of this superior academic identification attracted even more students. More buildings were constructed.

Ground was broken in April 1971 for another complex named in honor of Gill Robb Wilson, founder of the Civil Air Patrol and former editor of *Flying* magazine. The complex consisted of three pods, with the two-story structure later named the "Tine Davis Building," honoring one of the founders of the Winn-Dixie Food Stores, who was a benefactor of the University at times when financial support was an absolute necessity.

◀ **The Embry-Riddle administration building in 1972.**

Amid all this growth Jack Hunt needed solid leadership at the helm of the Board of Trustees. One Board member who possessed the necessary leadership qualities was Brig. General William W. Spruance. In 1972, Hunt approached the General with a request to serve as Chairman.

"I don't know. I've never done this before," answered Spruance.

"Then take it temporarily," said Hunt.

"I will, until you find someone who knows what he is doing," joked Spruance.

The General served as Chairman from that day until April 1987 when he was named "Chairman Emeritus." For his years of loyal service, the University voted in 1980 to name a house on the Prescott campus and, on October 23, 1987, a new administration building on the Daytona Beach campus in his honor.

During his 15-year term as Chairman, Spruance and his fellow Board members tackled their share of problems—some of which divided the members into separate camps.

One of the more enthusiastic debates came early in 1972 when the School searched for an appropriate home for its president.

Austin O. Combs, a community leader, Board member, and successful realtor, was never afraid to meet a challenge head-on, and negotiated the purchase of a large, ocean-front home on A-1-A. Combs faced plenty of opposition from some Board members. "Some of them thought this was too much of a financial undertaking," said Combs. "But we were able to convince a majority of them to purchase the site. Today, it's worth many times the purchase price."

► The President's Residence in 1995. The beach-front home was designed by well-known architect Fred Dana Marsh and the interior of the home displayed some of his original artwork. The residence served as a home for the President and his family and was also used for special University events, employee retreats, and other functions.

Jack Hunt saw the President's Residence in Daytona Beach (better known as "The Battleship" because of its unusual silhouette) as the setting for community activities and receptions. Even the most severe skeptic admits it was one of Embry-Riddle's wisest investments.

On the Campus, the ramp constructed for the University's airplane fleet was dedicated January 19, 1974, with a unique ribbon-cutting ceremony. Students and faculty watched in awe as 72-year-old John Paul Riddle, taxied a Cessna 172 through the ribbons onto the ramp.

It was a busy day for the participants who then walked to the site of the ground breaking for another new building—the "University Center."

Construction on the first of the three buildings of the Aviation Maintenance Technology Center began in 1975 and was available for classes in 1977. It was none too soon. The number of students continued to grow; fall enrollment for '77 reached an astounding 2,725.

The Wind Tunnel had to be relocated to make room for airport expansion. The Engineering Technology Building, a prefabricated metal structure, was completed in 1978 to consolidate the aeronautical engineering program and house the wind tunnel.

That same year, the second dormitory was under construction, and the Tine Davis/Winn-Dixie Swimming Pool was completed.

In addition to the growth of buildings and academic offerings, the University sponsored both Army and Air Force ROTC programs. Embry Riddle's Air Force ROTC eventually grew to become the largest volunteer detachment in the nation.

One of the main reasons why Embry-Riddle not only survived, but prospered during his 21 years as president, was the fact that Jack Hunt literally dedicated his life to make things work. Hunt was a bachelor during most of his presidency. That's not surprising to those who knew him. Embry-Riddle was his mistress, his passion, his family. He rarely took vacations. He waited until 1981 to marry Lynne, an administrator at the Prescott campus, and adopt her six children as his own.

In 13 years, the campus in Daytona Beach grew to a stature never imagined by those sweating, tired volunteers of "Operation Bootstrap" who stepped from their trucks after their nine-hour drive from Miami. Even the July 2, 1979, issue of *Time* magazine referred to Embry-Riddle as "the Harvard of the Sky." Certainly, no one would blame the Administration for taking time out for a few bows.

Some people, however, are never satisfied with the status quo. Even Jack Hunt's closest friends didn't realize that he was eyeing further opportunities for growth.

6

A World-Wide Impact

*"Embry-Riddle doesn't belong to me
or to the Board of Trustees.
It belongs to everybody in the
United States. And that, ladies
and gentlemen, is a big responsibility."*

— JACK R. HUNT, 1983

The Western Campus

"An opportunity such as this comes to us only once in a lifetime," said the visionary President Hunt to some members of the Board of Trustees in 1977. "Others consider it a lost cause," he continued. "For us, it's a chance to grow beyond all our former expectations."

Jack Hunt was talking about a large campus nine miles north of Prescott, Arizona, once called "Prescott College."

The college had a less than spectacular history. Once an institution that offered courses such as "Wilderness Survival" and was noted for its "Outward Bound" program, the school went bankrupt in 1974. Its students were told about the school's demise through a cursory announcement on December 18 and ordered to clear out by midnight. The campus lay deserted until 1977, when Hunt, a few Trustees, and some administrators walked the 511-acre campus near the Prescott Municipal Airport.

"Can you imagine a Western Campus for Embry-Riddle out here?" asked Hunt.

"It seems to me that you already have," answered Dr. Ronald Wiley, one of the University deans.

▲ Embry-Riddle's Prescott, Arizona campus in 1994.

Dr. Wiley, a psychologist, was correct in his analysis. At the persuasive urging of the president, the Board of Trustees voted in 1978 to purchase the property for a fraction of the assessed value.

The original plan was to establish a college preparatory school for eleventh and twelfth grade students, using flight as a motivation for academic performance. Within five years, the Trustees hoped to develop a four-year branch of the University at this site.

The college preparatory idea never gained acceptance by prospective students. Something else had to be done.

Jack Hunt and the Trustees acted quickly. The concept of a Western Campus was reclassified from "future plans" to "current project." Buildings were renovated; classrooms and office spaces were shaped from existing structures. By the fall of 1978, the campus opened its doors to 264 students and a small staff.

▶ **At the December 1994 Daytona Beach campus graduation ceremony, Dr. Sliwa *(far left)* welcomed visitors *(from left to right)* FAA Administrator David Hinson, Prince Fahad Bin Abdullah of Saudi Arabia, and Embry-Riddle Chairman of the Board of Trustees James G. O'Connor.**

"The students could smell the fresh paint and sawdust as they were registering for classes," said General Spruance, Chairman of the Board of Trustees, "but there was no mistake that we were here to stay."

Before champagne corks were popped in celebration, President Hunt warned of one possible danger: "The Prescott Campus is a part of the one university. Each part of the University is the same as a member of a family. We cannot entertain a plan that will in any way weaken the whole, or we will destroy what we have worked so hard to build."

The Prescott Campus that year offered just one degree—a Bachelor of Science in Aeronautical Science. The limited facilities and the hastily assembled student body combined to create a need for unique programming.

Students were divided into two groups—the gold and blue. The blue group attended academic classes on Monday, Wednesdays, and Fridays, and flight line activities on Tuesdays, Thursdays and Saturdays. The gold group alternated schedules for academics and flight.

Seven full-time and three part-time instructors carried the academic load. Twenty-six flight instructors taught in 15 Grumman "Tiger Cats."

Over the years, more buildings were remodeled. Student enrollments increased. The 1985 fall enrollment totaled 888; the staff numbered 160.

Paul Daly, then Chancellor of the Prescott Campus, worked diligently, leading the campus to grow both academically and numerically. "At first, we did not feel we were part of the University," he admitted. "We were dealing with dirt roads, someone else's school and its bad reputation. Because we entered the picture so late in its history, we felt like a step-child to the Daytona Beach campus. However, visits to the campus by J. Paul Riddle gave our students a touch with the past, the support of President Hunt and the leadership of Jeff Ledewitz and Bob Jost in terms of our budget planning, all helped to demonstrate that we were an important part of the Embry-Riddle family."

The Extended Campus

The investment into the Prescott Campus enabled Embry-Riddle to appeal to a greater segment of the nation. But if the University was going to reach students living outside the continental United States (particularly those serving at military bases), yet another approach had to be created.

Traditionally, Embry-Riddle attracted military personnel—especially those involved with some phase of aviation. It was natural, then, that the University find additional ways to serve this community. What the president and the Board of Trustees called "The International Campus" provided one approach that worked.

Resident centers were established as early as 1970 at military bases throughout the United States and Europe. The idea was to create an environment so that, without leaving the base, a member of the armed forces could attend classes, even complete all degree requirements for both the bachelor's and master's degrees. The idea worked. Today, through what is now known as the "Extended Campus," Embry-Riddle sponsors resident centers in over 100 locations, serving nearly 12,000 part-time students.

The International Campus of the early '70s reached out to another group of people who, because of careers and schedules, were unable to attend classes in a conventional campus environment. The Department for Independent Studies now allows nearly 1,000 men and women to study via self-teaching texts, individual counseling, and state-of-the-art, computer-aided programs.

Charles Williams, then chancellor of the International Campus, saw this as the campus with the greatest potential for meeting the needs of aviation professionals.

Because of increasing requests from graduates and other professionals in aviation, Embry-Riddle applied for and received

approval from SACS to offer a Master of Aviation Management degree during the summer of 1973. Arrangements were completed with Biscayne College and Barry College in Miami to offer the first program.

Jack Hunt saw Embry-Riddle as an exclusive school. Each student, he felt, who graduated from any of its campuses, should receive a degree that included the words: "aeronautical" or "aviation." In his opinion, "Unless Embry-Riddle is a unique institution of higher education, it has no reason to exist." At the same time, Hunt never connected "aviation" with "aerospace." "That's an entirely different element, and we have no business with programs that send people to the moon. We're here for support of the aviation industry, period."

Bunnell

Jack Hunt, who kept active his real estate license, was constantly on the lookout for property at a bargain price. One of those bargains was a lodge at a defunct amusement park in Bunnell, Florida, a rural community 20 miles north of the Daytona Beach Campus.

The "Marco Polo Lodge" at the intersection of Interstate 95 and the Old Dixie Highway was designed to house visitors to Marco Polo Park—a theme park located across the road. Unfortunately, no amusement center could compete with Walt Disney World in nearby Orlando. After operating for a bit more than one year, the executives of Marco Polo Park filed for bankruptcy. With the closing of the park, the lodge became virtually useless.

Mr. Jack White, one of the park's owners, approached Presi-

◄ Gathering at the 25th Anniversary Celebration are 10 volunteers who helped move Embry-Riddle from Miami to Daytona Beach on April 24–26, 1965. Shown from left to right are Beeman Alexander, Lou Fuchs, Floyd Tredway, Sully Ferrito, Wes Olson, Gary Cunningham, Agee Tacker (Embry-Riddle Aviation Safety Engineer and Professor), Alan Robertson, and Frank Forrest (former Embry-Riddle Adjunct Faculty Member).

dent Hunt with an offer to sell it to the University for the balance of the mortgage payments. In 1980 the agreement was signed.

The new Embry-Riddle "Main Campus" (as it was initially called) allowed certain departments to move from the over-crowded Daytona Beach Campus. The administrative offices and what was known as the "International Campus" were the first to relocate there. Others soon followed. By 1985, the "Executive Offices" (as it was renamed) housed the offices of the president and his staff, the International Campus, plus the departments that served all three campuses.

The Bunnell facility was used actively by the University until 1987 when construction of new buildings on the Daytona Beach campus allowed the operations to be shifted there.

Growth in Daytona

Not only was Embry-Riddle offering programs in response to the needs of the times, but also the "mother campus" at Daytona Beach had to change in order to accommodate the increased enrollment that sometimes surpassed 5,000 students.

Classroom buildings, apartment complexes, a service park, full-service library, swimming pool, and parking facilities were among the more notable additions.

In addition to single-engine and multi-engine aircraft on its flight line, Embry-Riddle adopted the use of simulators that gave students an opportunity to hone their skills in navigation and communication.

Simulators were only part of the advanced technology employed at the school. Students could earn an associate's and bachelor's degrees in avionics.

Three separate rooms house the digital, micro-processing, and communications laboratory, plus the linear-components and semi-conductor laboratory, and the avionics laboratory.

Although most of the growth was in the form of new buildings and more complex aircraft, the University insisted that each student be required to study English, history, philosophy, literature, psychology, sociology, and other similar disciplines. President Hunt insisted that aviation education should train the whole person. It was a principle not unlike that of J. Paul Riddle who, a half-century earlier, learned to appreciate the importance of all phases of aviation.

One of those phases, according to Hunt, did not involve high-brow professors who stood aloof from the real world. He was not impressed with overly sophisticated theorists ("acadamaniacs," he called them) of aviation or of any of the other subjects offered at the University. Hunt, instead, maintained that nobody can really teach; all he or she could hope to do was motivate students to learn. "One of the most powerful motivational appeals is aviation," he said.

A pragmatic proof that his philosophy was sound was the incorporation of the "Upward Bound" program. Beginning in 1975, selected students—those who were prime candidates to join the ranks of high school dropouts—assembled on the Daytona Beach Campus each Saturday morning and on weekdays for six weeks during the summer. Aviation was used as a context in which they learned reading, writing, and math. As a result, over 90 percent of its graduates are accepted into the colleges and universities of their choice.

One of the more remarkable changes on the Daytona Beach Campus was a decree from the Federal Aviation Administration in 1979 that granted self-examining authority for private, commercial, instrument, and multi-engine certificates. In 1983, the self-examining authority extended to license instructors in these areas. This unique exception meant that a student with high aptitude can save thousands of dollars in the search for advanced certificates.

This one pronouncement by the F.A.A. demonstrated more

than anything else that Embry-Riddle had matured to a level that must have surpassed even the wildest dreams of its leaders.

A New Era

On the morning of January 7, 1984, four ROTC cadets solemnly marched to the main flagpole on the Daytona Beach Campus. They raised the colors to the top, then slowly lowered them to half-mast. President Jack Hunt, age 65, died at his home following a year-long battle with cancer.

Many of the thousands who gathered for memorial services at the Daytona Beach and Prescott Campuses openly wept.

Jack Hunt was many things to many people. But one thing was certain. His life was the University, particularly the students.

"He always put the students first," said his widow, Lynne. "He turned Embry-Riddle from a near-bankrupt school with a few hundred students to a full-fledged university. And every decision he made was in light of the question: 'How will it help the student?'"

Jack Hunt's appreciation for the student became most apparent when he recommended to the Board of Trustees that students be represented as full voting members of the Board. The recommendation was approved and put into practice long before most other academic institutions ever considered such a step.

The student was paramount in terms of President Hunt's vision of quality education—a vision not always understood by accrediting agencies. He believed the University should offer the student a practical education that would help get the student a job and make him or her immediately productive in an aviation career. He was willing to debate anyone who equated quality education with the number of books on the shelf of a library. The fact that the library on the campus in Daytona Beach is named in his honor remains an irony to those who knew him.

While his standards may have challenged traditions, his com-

▲ The Embry-Riddle Aeronautical University campus in Daytona Beach, Florida in 1985.

▲ The Daytona campus in 1975. This aerial photograph shows the Daytona Beach campus buildings and other features.

mitment was no less real. He was less concerned with SAT scores of incoming students, instead he focused on the quality of graduates. "It may be easy to get into Embry-Riddle," he often said, "but it is not easy to get out."

In keeping with the theme begun by J. Paul Riddle in the 1920s, Hunt remained alert to the potential contribution of women to aviation and higher education. One of the employees, Dianne Thompson, began as a clerk/typist at the University in 1971. Later, she applied for the position as secretary to the president and was given the job.

"President Hunt recognized something in me I didn't know was there," she said. "He gave me an opportunity to grow and to develop a confidence I didn't know I had."

In 1976, with the urging of Hunt, Mrs. Thompson was elected corporate secretary of Embry-Riddle, making her an officer of the University.

The president remained alert to other values as well. Two months before he died, Jack Hunt was honored by the Daytona Beach community when the Regional Airport Terminal was named in his honor.

At the ceremony of dedication, a tired, thin, gravel-voiced Jack Hunt was still able to echo his vision of the University: "Embry-Riddle doesn't belong to me or to the Board of Trustees," he said. "It belongs to everybody in the United States. And that, ladies and gentlemen, is a big responsibility."

Shortly before January 7, when he knew the end was near, Jack Hunt requested that in lieu of purchasing flowers, donors give their contributions in his memory to the construction of the new library on the Daytona Beach Campus and to scholarships for deserving students. "Concern for the students was foremost in his mind—even in the face of death," said John Fidel, a vice president for the University.

Tributes to President Hunt poured in from throughout the

United States when an A.P. wire announced the news of his death. Aviators, educators, politicians, public figures, and perhaps most important, the University's students, praised his accomplishments.

Dr. Jeffrey H. Ledewitz, who took the reins as interim president for over a year, spoke for the entire Embry-Riddle family when he addressed the next meeting of the Board of Trustees:

> A university is a living thing. Fragile in infancy, it grows in strength as it matures. When guided with wisdom and vision in the formative years, it can develop a character and sense of purpose that will enable it to meet successfully the challenges of a changing world. Embry-Riddle Aeronautical Universitywas indeed fortunate to have such a leader in President Jack Hunt.

Dr. Ledewitz, who earned praises from the students, faculty, administration and community for his work as interim president, had openly expressed the fact that he did not wish to remain in the president's office on a permanent basis.

The search was on for someone with the skills and temperament to carry on the work begun by Jack Hunt.

7

The General
and His Mission

*"The biggest challenge in getting
the program in high gear was
to motivate everybody to work as a team.
I'm confident we did just that."*

— KENNETH G. TALLMAN

The General Arrives

The story of Embry-Riddle Aeronautical University is the result of a rare combination of dedicated pioneers, each of whom left a unique mark. People such as J. Paul Riddle, T. Higbee Embry, John McKay, Isabel McKay, and Jack Hunt were determined, individualistic, sometimes stubborn people who proved to be the right leaders for their times. In fact, it is hard for us to conceive of the University as we know it without the impact of their personalities.

In keeping with this tradition, the Board of Trustees at its meeting in January 1985, selected as the new president of the University, Kenneth L. Tallman.

Tallman, a native of Omaha, earned a solid reputation in the Air Force, rising to the rank of lieutenant general. As a graduate from the United States Military Academy and having received a master's from George Washington University, he brought with him an appreciation for academics. As an Air Force fighter pilot, commander, and personnel manager, he had solid firsthand experience with the military and aviation.

Aviation school administration was nothing new to the general. Prior to his Embry-Riddle appointment, he distinguished himself as superintendent of the United States Air Force Academy and as president of the Spartan School of Aeronautics.

General Tallman (he was more accustomed to the title "general" to that of "president") also brought to the University a philosophy of leadership that elicited enthusiastic cooperation from staff, faculty, and students alike. In some respects he was a behind-the-scenes manager. He shunned any hint of a dictatorial, "Do it my way or else" posture, and opted, instead, for listening to the advice of his staff. Although he sat at the top of the administrative totem pole, he was not above going out of his way to help students, employees or family members with personal problems.

General Tallman's management style was unlike that of some others who don the cap and gown of a university president or even his predecessors in many ways, yet he shared that one philosophy that has guided the School ever since Mr. Riddle first climbed into the back seat of a Waco-9: he continued to stress the fact that aviation is more than just flying airplanes.

Like his predecessors, he carried an appreciation for the entire scope of aviation, including air science, maintenance, engineering, management, computer science, avionics, meteorology, and, of course, safety.

This is not to imply that effective leadership at Embry-Riddle came easy for the three-star general. "When I was in the military and gave an order," he said, "it was carried out without hesitation. At Embry-Riddle, when I gave an order, many of the faculty and most of the administration demanded to know why. Some even suggested that we form a committee to assess the wisdom of my command. Now, that's a hard transition for anyone to make." But he did.

General Tallman adjusted his approach to leadership and was able to bring both faculty and staff in line with his goals. Those goals included a new emphasis in response to demands of the times. "Although Embry-Riddle intends to remain focused on avi-

▶ Lt. General Kenneth L. Tallman was named president of Embry-Riddle Aeronautical University in 1985.

ation," he often said, "it also plans to expand its curriculum to include aerospace." Tallman explained that aerospace is a future of aviation that many institutions have overlooked. "If we are going to make any sort of impact in the twenty-first century," he says, "we must lead the way in education for the age of aerospace. It's past the time that we think of aerospace as something apart from aviation."

Eric Doten, former chancellor of the Daytona Beach Campus, picks up on this theme: "We introduced an engineering physics degree program as an initial step into the aerospace arena. We saw this as an opportunity for future development in areas that we hadn't yet imagined."

To emphasize his point, General Tallman endorsed a change in the tradition soundly amplified by his predecessors. Instead of insisting that the words: "aeronautical" or "aviation" be included with each degree awarded a graduate, he permitted, even encouraged, the granting of degrees with more "conventional" titles.

General Tallman continued Jack Hunt's philosophy—"The student is our reason for being here"—but he accepted the more traditional academic standards of quality. Faculty academic credentials (and salaries) increased significantly during his tenure.

"It was also important that we not forget the value of research," added Tallman. "Even though we are primarily a teaching institution, we expanded our efforts in research in order to be more responsive to the aerospace industry, attracted grant monies and equipment, and provided development opportunities for our faculty." As a preliminary to expanded research activities, Embry-Riddle introduced a Master of Science Degree in Aeronautical Engineering commencing in the fall of 1986. Other opportunities for applied research were explored in the human factors engineering area.

◀ **This full-size replica of the Wright Flyer, constructed of stainless steel and aluminum and weighing around 4,000 pounds, has marked the entrance to Embry-Riddle's Daytona Beach campus since 1991.**

Expanding Embry-Riddle's Influence

General Tallman accepted as one of his challenges to let people know about "aviation's best kept secret" by publicizing the incredible growth and stature of Embry-Riddle. He was frequently called on to give Congressional testimony regarding aviation education and the need for research. He accompanied faculty and consultants to Egypt on an initiative that resulted in a consortium involving Embry-Riddle and the Egyptian National Civil Aviation Training Organization.

Embry-Riddle also grew and expanded through its International Campus (now the "Extended Campus") which services adult education needs. President Tallman pointed out that "this is a fast-growing market, with more and more people seeking opportunities to complete requirements for bachelor and master's degrees in aviation-oriented programs."

As an adjunct to this initiative, the General placed greater emphasis and resources into professional programs for the aviation-minded adult through workshops, seminars and short courses. "In this way," he explained, "we maintained curriculum relevance, faculty currency, and aviation industry rapport."

In March 1986, Embry-Riddle was awarded the management training and support services contract for the FAA. The center was constructed in Palm Coast, Florida, located approximately 30 miles north of the Daytona Beach Campus. "This contract was valued at $47 million with an economic impact of over $500 million," reported Dr. L. William Motzel, vice president for special projects, who was responsible for writing the proposal.

In that same month, the University received a letter from Donald D. Engen, administrator of the FAA, announcing the award of a $500,000 grant to begin development of an airway science simulation laboratory. "This gave the University a unique resource for helping students to understand the comprehensive dimensions of the national airspace system and allows them to develop judgment and team skills in that environment," said Dr. Motzel, who also wrote the proposal for this project.

"These two initiatives spearheaded by Dr. Motzel in support of FAA requirements provided some golden opportunities for greater interface with the aviation/aerospace industry, both nationally and internationally," said General Tallman. "They helped us continue to lead the way in aviation education as our air transportation system expands into the next century."

One of General Tallman's goals from the first day he set foot on campus was to move the administration building from rural Bunnell to Daytona Beach. Backed by his support and encouragement, the Board of Trustees voted to construct Spruance Hall in 1987 which today houses the offices of the president, his staff and other key personnel.

Baskets, Base Hits and Beautiful Landings

Other areas of emphasis and growth marked General Tallman's tenure as president. One of these was in the area of athletics. During its embryonic stages of development, when the University struggled to gain national recognition and to maintain a balanced budget, there was little time to think about organized sports. Outside of a spirited intramural program in which clubs and fraternities competed for bragging rights in volley ball and softball games, Embry-Riddle had not earned a reputation for intercollegiate sports.

That all changed in 1988 when "The Eagles" played its first intercollegiate basketball game with Hillsborough Community College. Embry-Riddle wound up on the short end of a 93–62 score. But the important fact was that a genuine intercollegiate program was on its way.

"We came from virtually nowhere," remembers Steve Ridder, basketball coach and director of athletics. "We had no facilities, no coaches, no teams, no scholarships and no uniforms. In fact,

the only inventory in our athletic department was a box of one dozen baseballs."

Since then, Ridder was named Central Florida College Coach of the Year and, on more than one occasion, Athletic Director of the Year.

Competitive matches in other sports—baseball, golf, soccer and tennis—were added to the University calendar. In 1995, women's volleyball began its intercollegiate competition.

In that same year, Embry-Riddle coach Greg Guilliams was selected as the NAIA Southeast Regional Coach of the Year. The Eagles finished 45 & 13 and ranked ninth in the final NAIA Top 25 poll.

Contests with other colleges and universities went beyond traditional sports. Embry-Riddle was gaining a reputation for its precision flight team in contests sponsored by the National Intercollegiate Flying Association. In fact, the Daytona Beach campus team won the national championship in 1992. Not to be outdone, the Prescott campus received a national championship trophy in 1993.

General Tallman's commitment to the education of the "whole person" included a spiritual dimension as well. He strongly supported Dr. James Plinton, a member of the Board of Trustees, in his initiative to establish an interfaith chapel in Daytona Beach which was completed in 1994 and included four prayer rooms (Catholic, Protestant, Jewish and Muslim) along with appropriate space for worship and counseling.

All of these accomplishments resulted not from the efforts of only one person, and General Tallman was quick to point that out. "The whole process of growth for Embry-Riddle was a multifaceted operation," he said, "and the biggest challenge was getting the program in high gear by motivating everybody to work as a team. I'm confident we did just that."

◄ The Daytona Beach campus Eagles basketball team won the 1993–94 Florida Sun Conference.

General Tallman's success as a leader in aviation higher education did not go unnoticed by others throughout the nation. On January 12, 1992, the General was awarded the coveted Frank G. Brewer Trophy—"in recognition of his 40 years of creative leadership, personal accomplishments and professional contributions to aviation and space education"—by the National Aeronautic Association. The trophy is on permanent display at the National Air and Space Museum in Washington, D.C.

Symbolic of the General's tenure as president was the construction of the Jack R. Hunt Memorial Library in Daytona Beach and the metal sculpture of the Wright "Flyer"—a monument suggested by the students as an appropriate "first sight" as students and visitors entered the campus. General Tallman personally raised the funds for the sculpture and guided its design and implementation.

On June 30, 1991, after six and a half years as president of Embry-Riddle, General Tallman retired to spend more time at home with his wife, Dee, and his grandchildren in Tallahassee, Florida.

The way was paved for another leader for Embry-Riddle.

8

Lift-Off to
the Future

*"It is relatively easy
to make a living;
it's much more important
to make a difference."*

— STEVEN M. SLIWA

Leadership Styles

Jack Hunt's wife and children remember Embry-Riddle's first president as a warm, sensitive husband and father who personified a gentle spirit of love and gentleness. That was a sharp contrast to his projected image while at the helm of the University. In Miami, Daytona Beach and Prescott, Hunt played the role of a brash, two-fisted, gutsy entrepreneur who dared to take the sort of risks that were needed to turn a tiny flight school into a nationally recognized university. This square- jawed California native approached his challenge as president in the same way as did a newly-appointed marshal who was called upon to clean up a town in the Wild West. To him it was not so much *how* something got done; of more importance was that it *got* done.

General Kenneth Tallman, on the other hand, used a more subtle approach. With the skills of one accustomed to military protocol, he, with the support of his wife Dee, chose to apply extra coats of polish and diplomacy that he felt were necessary to gain added support from both industry and the community. He led the University in developing a systematic approach to policies and procedures. He also brought the offices of administration back to the Daytona Beach Campus and insisted that the faculty begin to share in the governance of the institution. He recruited people to solve accreditation issues and he initiated intercolle-

giate sports. In short, Kenneth Tallman gave the University time to develop relationships and to mature.

No one can argue the point that both Jack Hunt and Kenneth Tallman were the right leaders for the right time. And either one would have been willing to adjust their management styles to fit the situation.

On July 1, 1991, a new kind of energy swept onto scene in the person of a 36-year-old entrepreneur—a risk-taker who dared to outline innovative strategies necessary to catapult the University into the 21st century.

Dr. Steven M. Sliwa was neither a military aviator (although he is a commercial pilot and certified flight instructor in both airplanes and gliders) nor a high-ranking military officer. Instead, he had earned his reputation in the fields of academics (a Ph.D. in engineering from Stanford) and industry (NASA research manager and vice president of an engineering software firm in Silicon Valley and founder of his own educational software company).

This is not to imply that Dr. Sliwa's nomination received a standing ovation from every member of the Board of Trustees. Some wanted to continue the tradition of a military-type leader akin to the recently-retired General Tallman. A few seriously questioned whether or not a 36-year-old man with no track record as a university administrator was the best choice at this time to lead Embry-Riddle. But the vast majority of the trustees realized that this was the man they needed to blaze new trails. "He is just the person we're looking for," said Dr. Edward Stimpson, chairman of the Board of Trustees and president of the General Aviation Manufacturers Association (GAMA). "We are excited about having him on board."

Friends of Embry-Riddle were actually surprised that Dr. Sliwa accepted the challenge as president for another reason. They knew that by taking this post, Dr. Sliwa would have to take a 70

◄ **Steven M. Sliwa, Ph.D., the president of Embry-Riddle since 1991, is shown taking the controls of one of the University's many aircraft.**

▲ The Prescott campus Eagles Precision Flight Team won the 1993 national NIFA/SAFECON competition.

▲ The Daytona Beach campus Eagles Precision Flight Team won the 1992 national NIFA/SAFECON competition.

percent cut in pay. They would soon learn his reason. "It's relatively easy to make a living," he told them, "but it's much more important to make a difference."

Dr. Sliwa saw his opportunity to make that difference in his role as the third president of Embry-Riddle Aeronautical University. It soon became clear that he, like his predecessors, was the right leader at the right place at the right time.

A Changing University

Dr. Sliwa joined the University just as a decline in aviation and aerospace enrollments hit the industry at the beginning of the recession of the early 1990s. He wasted little time in his attempts to move the University in a direction that would allow it to meet its academic and fiscal responsibilities. He openly spoke about his goals and about the necessary strategies that must be taken to reach those goals.

At first, not everyone appreciated Dr. Sliwa's "no nonsense" approach to solving problems. He had to cut the budget to close a $4 million deficit the second month after arriving. He reorganized the finances and guided the University to receive a $11.5 million appropriation from the United States Congress. These funds were used to launch Embry-Riddle into a $40 million capital expansion program. Operating funds could not be spent to balance the budget, but Dr. Sliwa convinced the Board of Trustees that it was important to make bold decisions in light of the challenges facing the ever-changing business cycle in the aviation and aerospace industry.

Shortly after he first sat at the president's desk, Dr. Sliwa dared to address some on-going problems facing the Prescott campus. "Prescott simply was losing too much money and enrollment was not up to expectations," he said. Consequently, he and members of the Board of Trustees seriously considered three options: 1) moving the campus to the recently vacated Williams Air Force Base; 2) closing the campus; 3) continuing operation of the campus with a fresh management approach.

Although no one could argue with the facts that led to Dr. Sliwa's questions, reaction from the Prescott faculty and community, along with many sympathetic voices, was supportive of maintaining the campus. The new president, along with the trustees, therefore, mapped out a five-year set of milestones aimed at making the Prescott campus financially solvent utilizing a more site-based management. The Prescott Campus immediately moved to meet these milestones. "After we worked out a plan toward a site-based management, Dr. Sliwa became a champion in support of Prescott's long-term viability," said Chancellor Paul Daly who served the campus from 1981 until his retirement in 1994.

▲ The new Prescott campus flightline.

▶ The use of advanced communications technology in the Instructional Center on the Daytona Beach campus allows people seated in the auditorium to watch events on a large screen that are occurring elsewhere, as well as accommodate distinguished lectures and speakers.

Under Daly's leadership, student enrollment increased from 500 to 1,700. In addition, he developed and implemented the "Campus 2000 Plan" which set as a goal a 2,000-student enrollment before the turn of the century. "Paul Daly proved to be instrumental in the progress and growth of the University in general and the Prescott campus specifically," said Dr. Sliwa.

In January 1995, a new chancellor—Dr. Stephen Kahne—was hired to fill the shoes of Paul Daly. Dr. Kahne, holding a Ph.D. in electrical engineering with an international reputation as a scholar and administrator, was welcomed by President Sliwa as "the ideal leader for the Prescott Campus as we face the challenges of the '90s."

According to Dr. Kahne, the Prescott campus offers a unique setting for an Embry-Riddle education. The small-town atmosphere provides a much different setting for both students and faculty than does a large city. "We see ourselves primarily as an undergraduate campus," said Dr. Kahne, "although we shall be participating in the graduate programs through the Extended Campus."

Dr. Kahne shares with President Sliwa a vision for the future.

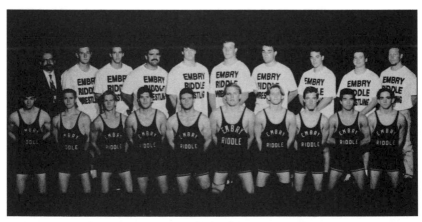

▲ The Prescott Wrestling Team ranked 21st in the nation after participating in the 1993–94 NAIA National Wrestling Tournament.

"Embry-Riddle has so much to offer," he says, "but it is relatively unknown outside of the aviation community. Every day we must work hard at transforming its image as a flight school and let the world know that we are a full-fledged university."

Accompanying Dr. Sliwa to the University was another strong source of influence in the community. His wife, Nancy (also a private pilot and former NASA program manager in robotics), immediately became active in service organizations and earned a solid reputation as a consultant. In addition, she trumpeted the cause of women in aviation by personally funding the Deanna Burt (an Embry-Riddle graduate) Scholarship for female students. "I wanted to give young women incentive to improve and assume responsibility in the maze of a masculine environment," she said.

Paul McDuffee, vice president of university relations, is one of the many faculty and staff who appreciates the vision of President Sliwa. McDuffee recalls his days as a student when the only library on campus consisted of a few shelves of books in a converted barracks and administration offices were cramped into structures that could barely be called buildings. He quickly admits that Dr. Sliwa does not fit the traditional mold of a college president. "He's a micro-burst of infectious energy who is not afraid to try things even if they're unconventional," he says.

"What is most obvious," says Steve Ridder—Embry-Riddle's award-winning basketball coach—"is the president's passion. He is consumed by the challenges facing him and the University."

Like those who preceded him, Dr. Sliwa's emphasis on student success is paramount. One of his first decisions was to name Dr. Jeff Ledewitz—former interim president and chief operating officer—as vice president of student life.

This decision received a standing ovation from both faculty and administration. Dr. Ledewitz—a gifted administrator—has always been at his best when working with undergraduates. He knew when to adopt a pastoral role when a student was overcome with grief or be as stern as any drill sergeant when admon-

▲ The Embry-Riddle Aeronautical University campus in Daytona Beach, Fla. in 1995.

ishing someone who stepped out of line.

From his first day on the job, Dr. Sliwa preached the gospel of team building. He surrounding himself with people who brought to the table a diversity of gifts. While President Sliwa articulated his visions, he encouraged talented people such as Paul McDuffee, Robert Jost, Jeffrey Ledewitz, Steven Kahne, Ira Jacobson and Christopher Mosher to utilize their talents and leadership skills in implementing those visions.

In order to help the administration keep its focus on the primary purpose of the University, Dr. Sliwa insisted that all academic administrators (including himself) teach a class.

Another of Dr. Sliwa's themes was that as the University matures, so do its people. According to him, alumni should be afforded every opportunity to feel proud of their Alma Mater.

▶ Volleyball is the first varsity sport for women at the University.

▲ The University Fieldhouse on the Daytona Beach campus consolidates
recreation and intramural activities, provides a home court for the men's
basketball and women's volleyball teams, and serves as a place to host
events and assemblies.

Both its buildings and the furnishings should reflect state-of-the-art technology. This included extended use of video conferencing where students in Prescott and Daytona Beach could simultaneously view lectures by professors and other experts in aviation. In addition, Dr. Sliwa and his staff initiated a dispensation of knowledge through Embry-Riddle's Aeronautical Virtual Library on the World Wide Web of the Internet. "What makes Embry-Riddle's web site unique is that it is the only one on the Internet dedicated solely to aviation and aerospace," said one administrator.

"Distance learning (the blending of telephone, television, and computer technology) is extremely important in this era of the rapid changes in education," said Dr. Ira Jacobson, vice president of academics. "The cost of on-campus education increases each year. Most of today's college graduates will expect to make several career changes during their lifetimes. Education no longer ends

▶ The King Engineering & Technology Center on the Prescott, Ariz. campus is
home to the electronics devises lab, the digital circuits and microcomputer
applications lab, the analog/digital control systems lab, and the power and
electronics lab, as well as the distance education center.

with the granting of a diploma; it is a life-long need. This new emphasis of Embry-Riddle caters to those of us in a mobile society.

"In this system, Daytona Beach students work on projects with students on the Prescott campus. This is a model for the real world."

With two major campuses and over 100 education centers at military bases throughout the world, Embry-Riddle is in a unique position to demonstrate creative solutions to modern-day challenges in education. With the new technology, Embry-Riddle can actually emulate the campus virtually anywhere in the world.

President Sliwa feels strongly that Embry-Riddle must adjust to the times. He expanded a practice begun by General Tallman by urging the University to award its graduates with even more degrees that did not include the words "aeronautical" or "aviation."

"It's a matter of pragmatic application to the needs of the industry," explains Dr. Sliwa. "There are more openings in the aviation and aerospace industry for those trained in a more general background than ones which are extremely specialized." Examples of new degree programs with broader applications but of significant importance to the aviation and aerospace industries include engineering physics, software engineering, technology management, and civil engineering.

Dr. Sliwa and the Board of Trustees envision Embry-Riddle as working hand-in-glove with government and industry in state-of-the-art research and development, balanced with a strong educational mission. President Sliwa and his team have been emphasizing partnerships with industry for developing training, cooperative education internships, and research programs for the mutual benefit of industry and Embry-Riddle. Specific areas

▶ **Prescott campus students redesigned and entered an experimental Indy-type electric racing car in the Arizona Public Service Electric 500 race. The race car, which has a top speed of 140mph, won third place in 1995.**

▲ **The Lehman Engineering and Technology Center groups five related academic departments and the Center for Aviation/Aerospace Research into an efficient teaching–research center.**

of concentration for research programs will be in human factors, advanced systems integration, manufacturing design, and aviation systems synthesis.

"The ideal environment for such progress," he says, "is one which combines aircraft, training simulators, research simulators, wind tunnels, computers and reference materials with academic programs that focus on developing, managing, maintaining and flying aircraft.

"Embry-Riddle is uniquely positioned to offer the aviation/ aerospace world tremendous support in all these areas."

Reaching Beyond the Box

Even to his early critics it became obvious that the new president not only was a visionary, but also was a man who had the intelligence to design workable strategies. In addition, he demonstrated the passion and energy necessary to motivate others to turn those visions into reality.

"Too many college presidents," according to Dr. Sliwa, "set self-imposed limits as to what they can or should do. They limit themselves to decisions and actions within the confines of the boxes they have drawn around themselves. Embry-Riddle is at the stage where it must reach beyond the box and discover its potential to serve the aviation/aerospace community.

"Some of that community lies within our neighborhood. Local school teachers, for example, come to the campus to get teaching materials, tapes, books, computer software and summer instruction programs that show how to use aviation as a method to teach subjects such as math, geometry and geography."

The aviation/aerospace community, of course, stretched beyond American borders. Foreign students who are required to know and use English—the international language of aviation—learn through the Embry-Riddle Language Institute (ERLI). In the summer of 1995, 30 air traffic controllers from China spent seven weeks at the Daytona Beach Campus learning how to direct pilots in clear English. This opened the doors to shorter routes. "As a result of this one program," said a spokesman for Delta Airlines, "during certain times of the year, we can save one fuel stop and from 35 minutes to more than an hour of flying time. That translates into a savings of $5,000 to $15,000 for each flight."

Recognizing the importance of the international impact of aviation, Embry-Riddle awarded an honorary doctorate in 1994 to His Royal Highness Prince Fahad Bin Abdullah in recognition of his accomplishments in military and civilian aviation in Saudi Arabia. Prince Fahad is in charge of civil aviation, which includes overseeing operation of the national airline, the nation's airports and civil aviation throughout the Saudi kingdom.

Demonstrating that Embry-Riddle was serious about its commitment to enhancing student life and enrichment, the long-awaited University Fieldhouse in Daytona Beach opened its doors in late summer, 1995. Now, Embry-Riddle can host student, alumni and local fans at sporting events and assemblies.

The Center for Aviation/Aerospace Research (CAAR) became the University's research operating arm. Its primary thrust is to enhance aviation/aerospace safety and to improve the effectiveness and efficiency of the National Airspace System through the development of airway systems safety technology, improvements to air traffic control, innovative flight technology, cockpit resource management, and related human factors gains.

As a testimony to the growing respect for its potential contribution to the aviation/aerospace community, in 1992 Embry-Riddle was awarded an $11.5-million capital appropriation from the Federal Aviation Administration for construction of its bi-located engineering buildings: the Lehman Engineering and Technology Center in Daytona Beach and the King Engineering and Technology Center in Prescott.

The Prescott facility added an activities center, the Robertson Aviation Safety Center and a simulator center to house Boeing 727 Level A simulators.

Sophisticated simulators with virtual reality technology have, for years, provided the atmosphere for training and certification of pilots for commercial airlines. Embry-Riddle adopted that technology in grand style in September 1995 when the University signed an agreement with FlightSafety International for a $25 million simulator center on the Daytona Beach campus that would house FAA Level D simulators for Boeing 737-300 and Beech 1900D airliners. A. L. Ueltschi, president of FlightSafety, predicted that through this state-of-the-art training, an Embry-Riddle aeronautical science graduate will be more marketable because of the realistic flying experience.

▶ Shown at right is a prototype of a FAA Level 'D' simulator built by FlightSafety International for Embry-Riddle Aeronautical University's Advanced Flight Simulation Center in Daytona Beach.

Alumni who attended the University during its formative years would hardly recognize their Alma Mater today. Since President Sliwa's arrival, both campuses sparkle with new state-of-the-art buildings. In addition, Dr. Sliwa has introduced new programs that will improve the value of a graduate's degree.

As they prepare for an even brighter future, those who embody the true spirit of the University will say: "We can do this ...and more."

And why not? For others, the sky may be the limit; for Embry-Riddle, it's home.

Epilogue

"Mr. Riddle's Last Flight"

On April 6, 1989, following a short illness, John Paul Riddle died at his home in Coral Gables, Florida. He was 87 years young.

Throughout his life he maintained contact with the school he began 63 years earlier in Cincinnati. When Mr. Riddle visited the campus, he often stayed overnight in one of the dorms and enjoyed walking around campus, meeting students.

"He was a dynamic guy who believed that there's no room for less than 100 percent professionalism," said student Mike Zaccaria. "He's the founder of aviation education as far as I'm concerned."

Kelli Young, another student, fondly remembered the times Mr. Riddle would join her and other students at a pizza parlor and talk about the early days of aviation. Other students testify that even when he had passed his 80th birthday, Riddle remained a worthy opponent on the tennis court.

The pioneer aviator once described flying as "the joy of being up in the clouds, free as a bird." His last airplane ride was on April 10 when, following a funeral service in Coral Gables and a memorial service at the University Center, his ashes were sprinkled over Biscayne Bay and in the Oak Ridge Cemetery at Arcadia, Florida.

▶ J. Paul Riddlle, co-founder of Embry-Riddle, died in Coral Gables, Fla., on April 6, 1989. A frequent visitor to both the Daytona and Prescott campuses, his greatest joy was "hanging out with the students."

Index

Interested in attending Embry-Riddle?

Contact Embry-Riddle Aeronautical University at any one of the following locations:

Daytona Beach Campus
600 S. Clyde Morris Blvd.
Daytona Beach, FL 32114-3900
(904) 226-6000 or (800) 222-3728
URL:http://www.db.erau.edu
E-mail (admissions): admit@db.erau.edu

Prescott Campus
3200 Willow Creek Road
Prescott, AZ 86301-3720
(520) 776-3728 or (800) 888-3728
URL:http://www.pr.erau.edu
E-mail (admissions): admit@ pr.erau.edu

Extended Campus
600 S. Clyde Morris Blvd.
Daytona Beach, FL 32114-3900
(904) 226-6910 or (800) 522-6787
URL:http://ec.db.erau.edu
E-mail (admissions): ecinfo@ec.db.erau.edu
Department of Independent Studies:
Graduate (800) 866-6271
Undergraduate (800) 359-3728

In Europe Contact:
Unit 4495
APO 09094-4495
Telephone from U.S.: 011-49-631-536-7170/7152
Military Telephone: 489-7170/7152
Voice Mail from U.S.: 011-49-631-536-7226

About John McCollister

"John McCollister is an incurable romantic," wrote the reviewer of one of his books. Perhaps that best describes the attitude that has resulted in his varied career. In addition to writing hundreds of newspaper and magazine articles and 13 published books, he has been a professor, university administrator, federal arbitrator, parish pastor and, of course, private pilot.

He was graduated with a B.A. in History from Capital University. Later he received his Masters in Divinity from Trinity Lutheran Seminary in Columbus, Ohio, and his Ph.D. in Communications from Michigan State University.

Dr. McCollister has been associated with Embry-Riddle since 1976 when he taught a variety of courses in the department of humanities. He also served the University as consultant to the president, director of development and director of professional programs.

He has earned his living as a writer, teacher and speaker, yet his passion has always been aviation. He is the owner of a classic Ercoupe which he flies every chance he gets from his home at the Spruce Creek Fly-in, Daytona Beach, Florida.

About Diann Davis

Diann Davis, formerly Diann Ramsden, served in various positions at Embry-Riddle Aeronautical University (ERAU) from 1979–1987, including Interim Director of the Learning Resources Center (now the Jack R. Hunt Memorial Library). She served as the faculty advisor for the charter student chapter of the 99s International Women Pilot Association and was instrumental in the creation of Embry-Riddle's first women's organization, Future Professional Women in Aviation.

While at Embry-Riddle, Ms. Davis created the Embry-Riddle Archives and served as the University Historian. She conducted customized presentations on Embry-Riddle's history for campus and civic organizations. Sparked by the colorful history of ERAU and personal friendship with J. Paul Riddle, Ms. Davis began documenting the history of ERAU using the Embry-Riddle archives. She visited the Cincinnati Historical Society, the newspaper department of the Hamilton County (Ohio) Library, and Lunken Airport to document the founding and early history of Embry-Riddle.

Upon leaving ERAU, Ms. Davis served in various positions at CLSI, Inc., a library automation provider. In 1991, she became the Coordinator of Training and User Support for the College Center for Library Automation in Tallahassee which connects Florida's 28 community colleges and their associated 59 campuses. She continues to follow the successes of Embry-Riddle through her many friends, colleagues, and professional contacts there.

Ms. Davis received her bachelor's degree from the University of Central Florida in Orlando and a masters' degree in Library and Information Science from the University of South Florida in Tampa.

LIVINGSTONE MOUSE

by Pamela Duncan Edwards
illustrated by Henry Cole

HarperCollinsPublishers

Livingstone Mouse was an explorer.

He spent his days running here and there, investigating this and that.

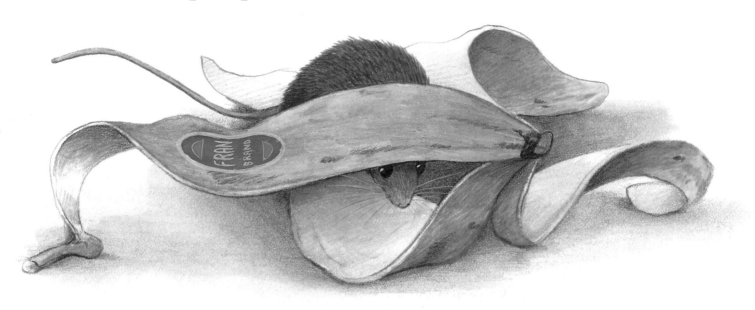

One night, Livingstone's mother said to her little mice, "You've grown too big to live in my small nest. It's time for you to find places to build nests of your own."

"I'd like to build my nest in the greatest place in the world," cried Livingstone. "Mother, where is that?"

"Well," replied his mother, "I have heard that China is very nice."

"That's it," announced Livingstone, kissing his mother good-bye. "I will go to find China."

After a while, Livingstone came
to the biggest building he had
ever seen.

"It's already divided into nest
places," said Livingstone. "That's
incredible. This must be China."

6

7

"There's lots of room," murmured Livingstone. He sat down to plan his nest.

A loud clicking noise came from next door. Livingstone saw a beetle turning somersaults.

"Excuse me," said Livingstone, "but do you do that often?"

"All the time," replied the beetle. "I'm a click beetle."

Suddenly, an angry voice yelled, "Stay over on your own side."

"Stay over on your own side yourself," yelled a second voice. "I don't know why I live with you—you never stop complaining."

"Excuse me," said Livingstone, "but do they do that often?"

"All the time," replied the beetle. "Cockroaches are so argumentative."

"I had no idea China would be so noisy," cried Livingstone.

"China?" laughed the beetle. "This isn't China. Why don't you try that way?"

9

Livingstone scampered
in the direction the
beetle had pointed. He
found himself in front of
a tall, white shape rising
majestically into the air.
Up one of its slopes ran
a rope ladder.

"That's fantastic," said
Livingstone. "This must
be China."

He ran up the rope ladder and down into the dark interior.

"It's nice and quiet," murmured Livingstone. He sat down to plan his nest.

A nasty, musty, human smell hit Livingstone's nose.

"Ugh!" he gasped. "I had no idea China would smell so bad."

"China?" chittered a camel cricket. "This isn't China. Why don't you try that way?"

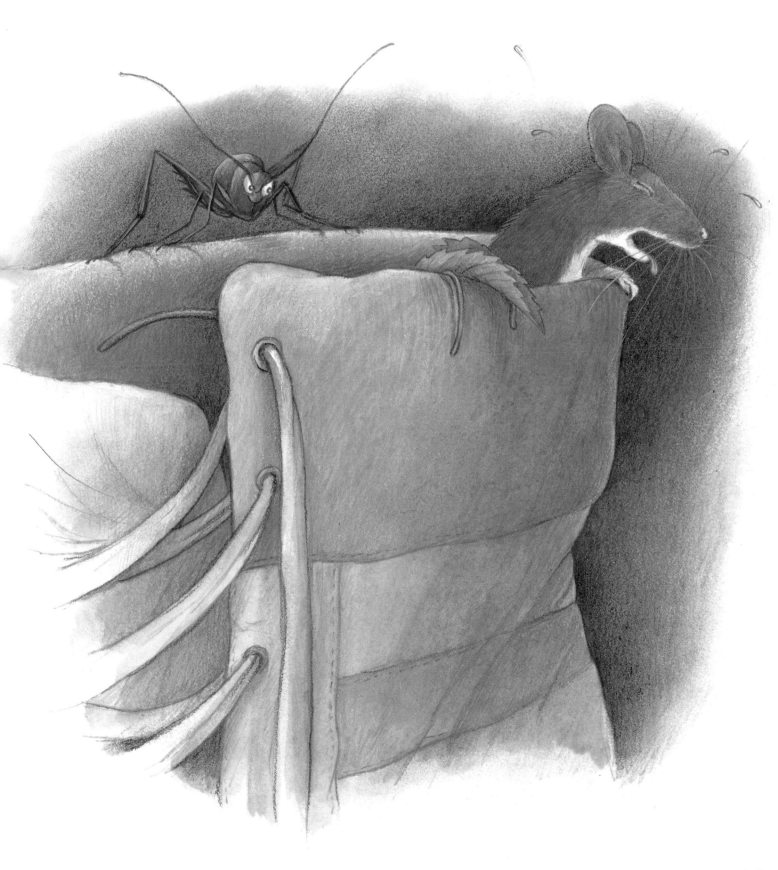

Livingstone ran on, gulping in air to get rid of the nasty smell. He skittered to a halt before a steep slope made out of pieces of thin, woven wood. From the top of the slope, Livingstone could see peculiar objects scattered about a cloth.

"That's wonderful," said Livingstone. "This must be China."

15

With a leap, he landed on the cloth. "It's quiet, and there's no bad smell," murmured Livingstone. He sat down to plan his nest.

Livingstone noticed a bottle with a big drop of red liquid dripping from its open neck. He took a big mouthful.

"Yeeow!" he howled. "I had no idea food would be so hot in China."

"China?" barked a raccoon, on his way to the trash cans. "This isn't China. Why don't you try that way?"

Livingstone plodded off, his tongue burning. Soon, he came to a square room with an open gate at the entrance. In the middle of the room, a piece of cheese lay on top of a slope.

"That's neat," said Livingstone. "It's quiet, there's no bad smell, and I'm sure that's mouse food. This must be China." He sat down to plan his nest.

Livingstone's mouth began to water at the thought of the cheese. "I'll eat a tiny piece," he thought.

"Run!" someone squealed, so loudly the cheese began to wobble.

"Eek!" screeched Livingstone as the gate clanked down, trapping the end of his tail.

"Dear, dear, dear," fussed the voice. "That was a silly thing to do."

Two large rats heaved at the bottom of the gate.

"You're lucky," said one of the rats, holding up his tail. "Look what happened to me."

"I had no idea living in China would hurt so much," cried Livingstone.

"China?" laughed the other rat. "This isn't China."

"Why don't you try that way?" they said together.

Livingstone scurried across the damp grass, shuddering to think what might have happened to his tail. Eventually, he came to a tall wooden post. Sitting on top of the post was the perfect little house for a mouse.

"That's amazing," said Livingstone. "I love the pointed roof and big picture windows. This must be China."

23

"It's quiet," murmured Livingstone. "There's no bad smell, no food to sting my tongue, and no mean gate to hurt me." He sat down to plan his nest.

With a loud pop, light flooded the little house, blinding Livingstone. Livingstone wiped tears from his eyes.

"I had no idea nights would be so bright in China," he sighed to a luna moth.

"China?" gulped the moth shyly. "This isn't China. Why don't you try that way?"

25

Poor Livingstone. He had a headache from too much
noise, he felt sick from the bad smell, his tongue
smarted, his tail throbbed, his eyes were
stinging, and he still hadn't found
China. Wearily, he set off again.
Then he saw it.

It lay on its side in a
shaft of moonlight, glistening
blues and whites. It had a bridge
to enter by, and a large inner room
with a secret passage leading out on one side.
"Oh!" gasped Livingstone, hardly daring to breathe.
"That is the GREATEST."

27

A bat glided over Livingstone's head.

"Excuse me," called Livingstone, "could you tell me what that great place is?"

"Oh, that," sniffed the bat. "You don't want to bother with that. That's just some old china."

"Some old China! But I've been looking everywhere for China," cried Livingstone. "This is where I want to build my nest and live happily ever after."

And that's exactly what he did.

For Geoffrey, who <u>is</u> the greatest.
—P.D.E
To Fran, explorer and wonderful friend.
—H.C.

Livingstone Mouse
Text copyright © 1996 by Pamela Duncan Edwards
Illustrations copyright © 1996 by Henry Cole
Printed in the U.S.A. All rights reserved.

Library of Congress Cataloging-in-Publication Data
Edwards, Pamela.
  Livingstone Mouse / by Pamela Duncan Edwards ; illustrated by Henry Cole.
    p.   cm.
  Summary: An explorer mouse in search of China discovers that he must be
careful to choose a new home that does not offend his sense of smell, sight,
hearing, taste, or touch.
    ISBN 0-06-025869-1. — ISBN 0-06-025870-5 (lib. bdg.)
    [1. Mice—Fiction.   2. Dwellings—Fiction.]   I. Cole, Henry, ill.   II. Title.
PZ7.E26365Li   1996   95-19981
[E]—dc20   CIP
                                                                              AC

Typography by Elynn Cohen
2   3   4   5   6   7   8   9   10
❖